T0413458

Automatic Calibration and Reconstruction for Active Vision Systems

International Series on
INTELLIGENT SYSTEMS, CONTROL, AND AUTOMATION:
SCIENCE AND ENGINEERING

VOLUME 57

For further volumes:
www.springer.com/series/6259

Beiwei Zhang • Y. F. Li

Automatic Calibration and Reconstruction for Active Vision Systems

Beiwei Zhang
Nanjing University of Finance and Economics
Wen Yuan Road 3
210046 Nanjing
People's Republic of China
e-mail: zhangbeiwei@163.com

Prof. Y. F. Li
Department of Manufacturing Engineering and Engineering Management
City University of Hong Kong
Tat Chee Avenue 83
Hong Kong
People's Republic of China
e-mail: meyfli@cityu.edu.hk

ISBN 978-94-007-2653-6 e-ISBN 978-94-007-2654-3
DOI 10.1007/978-94-007-2654-3
Springer Dordrecht Heidelberg London New York

Library of Congress Control Number: 2011941748

Printed on acid-free paper

Springer is part of Springer Science+Business Media (www.springer.com)

Preface

The research of computer vision is a systematic and inter-discipline science. And one of the most important goals is to endow a vision system with the ability to sense and cognize its environment from the two-dimensional image or image sequences. The critical step towards that goal is to build an appropriate imaging model for the system and estimate its parameters as accurately as possible, which is generally termed as camera calibration.

In practice, the image data are infected by various noises. Furthermore, the sensor model itself is generally obtained with some kind of approximation. As a result, it is difficult to realize an accurate, reliable and robust vision system due to the error accumulations. In the past decades, many distinct techniques have been suggested to cope with this problem, making use of either different physical features or geometric constraints, such as coplanarity, orthogonality and collinearity, etc.

The purpose of this book is to present our work within the field of camera calibration and 3D reconstruction. Several possible solutions for the related problems are discussed. Here, various cues are employed, including different planar patterns, plane-based homographic and fundamental matrix. The involved system ranges from passive vision system to active vision system, including structured light system and catadioptric system.

We thanks for so many people for their kindly suggestions and generous helps during the research work. They are Prof. FC Wu, Prof. YH Wu, Prof. S Y Chen, Prof. HN Xuan, Dr. Y Li and Dr. WM Li, etc. Special thanks go to the Springer editors for their efforts on this book. The work was financially supported by the Natural Science Foundation of China and the Research Grants Council of Hong Kong.

NanJing, China Beiwei Zhang

Contents

Chapter 1
Introduction

An efficient vision system is important for a robot to see and behave in an office or structured environments. Automatic calibration and 3D reconstruction are the critical steps towards those vision tasks. In this chapter, we firstly give a brief introduction about the background of the vision system and a general review of the relevant work. Then we present the scope, objectives and motivations of this book.

1.1 Vision Framework

Since 1960s, many researchers in the vision community have been studying ways and models to endow a computer or a robot with the ability of 'seeing', by imitating the animal or human vision system [1]. In 1982, Marr [2] described a well-known bottom-up computational framework for the computer vision system, in which three principal stages were considered from an information processing point of view. Firstly, the primal features from the raw images were obtained by edge detection and image segmentation. In this stage, various image features (such as corner points, line segments and edges) are detected, extracted and post-processed (such as establishing correspondences). Next, the structures and surface orientations were computed by applying the available constraints, such as shading/shadow, texture, motion and stereo. Finally, the high-level 3-D model representation and understanding were obtained. In this framework, each stage may be considered as an independent model with distinct functions, and divided into several sub-problems. For example, in the second stage, it mainly deals with the problems of camera calibration, stereo vision and shape-from-motion, etc.

With Marr's computational framework, different kinds of vision system have been developed. Generally speaking, they can be classified into several families following different criterions. For example, we can have passive vision and active vision system depending on whether there is energy (e.g. laser point, laser plane or structured light) projected into the scene intentionally. In turn, the passive vision system is respectively termed as monocular, stereo or trinocular system depending on the number of cameras involved. According to the optical configuration, we can

B. Zhang, Y. F. Li, *Automatic Calibration and Reconstruction for Active Vision Systems*,
Intelligent Systems, Control and Automation: Science and Engineering,
DOI 10.1007/978-94-007-2654-3_1,

Table 1.1 Camera calibration under various cues

Cues	Categories	Representative work
Dimensions	0-dimension (self calibration)	Pollefeys et al. [41]
	1-dimension (line-based method)	Zhang [175]
	2-dimension (plane-based method)	Zhang [34]
	3-dimension	Wilczkowiak et al. [33]
Motions	Static camera and static objects	Abdel-Aziz [87]
	Static camera and moving objects	Xie [135]
	Moving camera and static objects	Wang [169]
	Moving camera and moving objects	

have conventional vision system and catadioptric vision system (combining a CCD camera with curved mirrors). This indicates that the study of computer vision is multifaceted. Hence, the computer vision is an inter-disciplinary science, involving mathematics, projective geometry, computer graphics and physics, etc.

In this book, we concentrate on the second stage within the Marr's framework, i.e. problems of camera calibration and 3D reconstruction. In practice, some environments may be textureless and the active vision system with structured lighting is preferable. And the catadioptric vision system with larger field of view is required for some applications, e.g. visual surveillance and navigation. The former system is the combination of a DLP projector and a CCD camera while the latter integrates a specially shaped mirror and the CCD camera. Hence, we can name them the hybrid vision system in this work.

1.2 Background

The vision system plays the role of a bridge relating the 3D world and its 2D image, in which the main task of camera calibration is to estimate the intrinsic as well as the extrinsic parameters of the system. Essentially, it is to find the mapping from 3D light rays to 2D image pixels, or equivalently find a best model that fits those data [173]. A great deal of affords have been spent on the effective solution of this problem during the past few decades. Table 1.1 provides a rough overview of the existing papers following different cues.

The 3D reconstruction goes closely along with the camera calibration in computer vision, which deals with the problem of recovering the structure of an object or 3D information of the environment from its 2D image or image sequence (Fig. 1.1). It can be realized through the techniques of structure from motion, shape from shading and triangulation operation, etc. In fact, there is no distinct difference between those problems and they are different faces of the same problem. For example, the determination of object points is always performed with respect to some coordinate system. In this sense, the displacements are nothing but changes in coordinates. In what follows, we will give a survey on the recent development in both calibrated reconstruction and uncalibrated reconstruction.

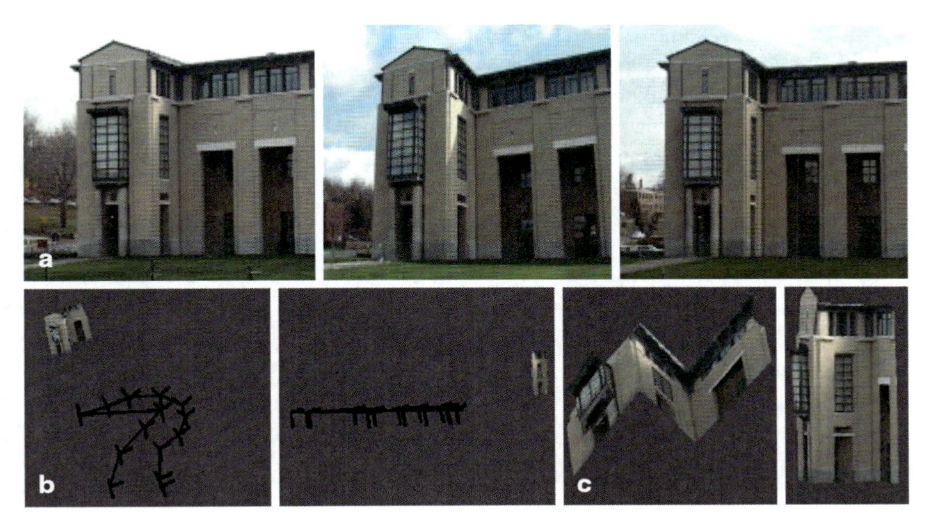

Fig. 1.1 An example of the 3D reconstruction from Han [174]

1.2.1 Calibrated Reconstruction

In this kind of techniques, partial or full camera calibration is required which directly results in a metric or Euclidean reconstruction, since the calibration provides some metric information about the scene [58, 59]. Depending on whether the vision system can be adjusted during vision tasks and whether a special calibration object is required or not, the methods can be roughly divided into two categories, i.e. static calibration and dynamic calibration.

1.2.1.1 Static Calibration Based Methods

In static calibration, a calibration target or device is elaborately designed with special structure to provide an array of points whose world coordinates are precisely known. Given a set of 3D points in general position and their corresponding points in the images, all the parameters of the system can be obtained by the maximum likelihood estimation method (MLE). Usually, there is a compromise here: the system can be accurately modeled with more parameters, but the coupling parameters increase the computational complexity and decrease the estimation precision at the same time. One of the most popular static calibration techniques is developed by Tsai [30]. This two-step method relies on the availability of a 3-D calibration object with special markers on it. In addition, it is required that the markers are not all coplanar. Since then, various calibration targets have been tried such as ellipses [31], planar curves [32] and Parallelepipeds [33].

However, the accuracy of the calibration results depends heavily on the characteristics of the target given a system model. A good calibration can only be assured if the feature points are accurately known and distributed inside the entire working volume. However, in a larger working volume, an adequate 3D object is often difficult to manipulate and expensive to produce with the accuracy needed for a reliable calibration. Therefore, there has been considerable research into reducing the level of metric information required without compromising accuracy. In particular, accurate planar objects are considerably easier to obtain than accurate regular solids. In 2000, Zhang proposed a flexible technique is for camera calibration [34], where a planar pattern with known structured information is involved. When calibrating, either the pattern or the camera is required to be placed at a few (at least two) different locations. Then a closed form solution can be obtained followed by a nonlinear refinement based on the maximum likelihood criterion. Sturm and Maybank [35] extended this technique by allowing possible variable intrinsic parameters. This algorithm can deal with arbitrary numbers of views and calibration planes and easily incorporate the known values of intrinsic parameters. In [36], Ueshiba and Tomita find that, a problem arises when applying Zhang's method to multi-camera system by simply repeating the calibration procedure for each camera: the rigid transformation between the pairs of cameras may be inconsistent in the presence of noise since the calibration is performed for each camera independently. To address this shortcoming, they employ the factorization-based method to construct a measurement matrix by stacking Homography matrices and then decompose it into the camera projections and the plane parameters. One the other hand, Shashua and Avidan [37] and Manor and Irani [38] independently arrived at the same conclusion that the collection of Homography matrices spans a 4-dimensional linear subspace. The difference lies in that the former studied the Homographies of multiple planes between a pair of images, while the latter deals with a pair of planes over multiple images as well as multiple planes across multiple images. In 2002, Malis and Cipolla [39] also presented their contribution on the multi-view geometry of multiple planes by constructing a super-collineation and super-E-collineation matrix.

1.2.1.2 Dynamic Calibration Based Methods

In some circumstances, it is very inconvenient or impossible to place a pre-made calibration object in static calibration. For example, in a video indexing application, we are only given the final video data without knowing what type of video camera these videos were taken from, not to mention getting the same camera to take a video sequence of the calibration object. Furthermore, the intrinsic parameters of the camera may be changing during the acquisition of the video sequence because of focusing and zooming. Therefore, 3-D reconstruction tasks under these scenarios would have to be performed using online calibration methods. The key difference between the online calibrated reconstruction methods is the way the camera's parameters are estimated. This on-the-fly estimation of camera parameters is often referred to as camera self-calibration or dynamic calibration.

Absolute-Conic Based Methods The absolute conic and its projection in the images play an important role in the process of camera self-calibration and hence Euclidean reconstruction. It is a degenerate point conic on the plane at infinity. As one of the earliest work, Maybank and Faugeras [40] employed the absolute conic for the auto-calibration tasks, where both the intrinsic parameters and metric scene reconstruction were obtained from multiple images. Pollefeys et al. [41] proposed a self-calibration method using the absolute quadric by assuming varying intrinsic parameters, where a two-view calibration problem was analyzed revealing that the effective focal length of each image could be obtained by solving a fourth degree polynomial equation. Triggs [42] presented a technique for camera auto-calibration and scaled Euclidean structure and motion from three or more views taken by a moving camera with fixed but unknown intrinsic parameters. One of the problems here is that, the method is applicable only if there are sufficient rotations about at least two non-parallel axes in the images and some translations in each direction, otherwise a poor estimation of the 3D structure will be caused. To overcome this drawback, Hassanpour and Atalay [43] modified Triggs' algorithm to incorporate known aspect ratio and skew values to make it for small rotation around a single axis. Besides these techniques, the absolute quadric is used for 3D reconstruction in [44] and [45]. On the other hand, the Kruppa coefficient matrix in the Kruppa equation, representing the dual image of the absolute conic, is also explored for self-calibration. By singular value decomposition of the fundamental matrix, Hartley [46] and Manolis [47] independently revealed their relationship and explored the applications. The earliest self-calibration method using Kruppa equations was proposed by Faugeras et al. [48]. However, when the motion of the camera is purely translational, the Kruppa equations reduce to a tautology. The ambiguities of calibration by Kruppa equations are discussed in [49].

Planar Homography Based Methods A method for 3D reconstruction based on Homography mapping from calibrated stereo systems was described in [50]. Here, the authors investigated the problem of recovering the scene metric structure by mapping one image into the other using Homographies induced by planar surfaces in the scene. Using at least four coplanar correspondences, the 3D structure can be achieved in Euclidean space up to a scale factor (scaled Euclidean structure) and two real solutions. In order to disambiguate the two solutions, a third view is required. Liebowitz and Zisserman [51] proposed an algorithm of metric rectification for perspective images using metric information such as a known angle, two equal though unknown angles and a known length ratio. They also mentioned the calibration of the internal camera parameters was possible provided at least three such rectified planes. Considering as a dual case, some researchers have addressed the problem of planar motion. For example, Knight et al. [52] showed that, the complex eigenvectors of the plane-based Homography are coincident with the circular points of the motion, when a camera undergoing a motion about a plane-normal rotation axis. Hence, three such Homographies provide sufficient information to solve for the image of the absolute conic and then the calibration parameters. Quan et al. [53] studied the constrained planar motion where the orientation of the camera was known. They demonstrated that the image geometry became greatly simplified by decomposing the 2D image into two 1D images: one 1D projective image and one 1D affine. Therefore, the 3D reconstruction was decomposed into the reconstruction in two subspaces: a 2D metric reconstruction and a 1D affine reconstruction. Gurdjos et al. [54–57] gave an extensive investigation on the problem of camera self-calibration by plane-based Homography from both algebraic and geometric viewpoints.

The above techniques are mainly developed based on the passive vision system. For the active vision systems, most existing methods are static and manual calibration. The calibration target usually needs to be placed at several accurately known or measured positions in front of the system [70–72]. For example, Bouguet [73] and Fisher [74] independently designed a weak structured light to extract the 3D shape of an object. In [75] and [76], the authors investigated a technique for calibrating their range finder systems which consisted of a matrix of 361 laser rays. The major limitation is that the system cannot be used for large depth measurement [77] and [78] presented extensive survey on the existing work. With these traditional methods, the system must be calibrated again if the vision sensor is moved or the relative pose between the camera and the projector is changed. However, only a few related works can be found on dynamic calibration of a structured-light system. Among the existing works, a self-reference method [79] was proposed by Hebert to avoid using the external calibrating device and manual operations. Fofi et al. discussed the problem in self-calibrating a structured light sensor [80]. However, the work was based on the assumption that "projecting a square onto a planar surface, the more generic quadrilateral formed onto the surface is a parallelogram". This assumption is questionable. Considering an inclined plane placed in front of the camera and the projector, projecting a square on it forms an irregular quadrangle instead of parallelogram as the two line segments will have different lengths on the image plane due to their different distances to the sensor. Jokinen's method [81] of self-calibration of light stripe systems is based on multiple views. The limitation of this method is that it requires a special device to hold and move the object. Recently, Koninckx [101, 102] presented a self-adaptive system in which 3D Reconstruction was based on a single frame structured light illumination.

1.2.1.3 Relative Pose Problem

In general, the relative pose estimation solves the following problem: given a set of tracked 2D image features or correspondences between 3D and 2D features, finding the extrinsic parameters of the camera in a calibrated vision system. It is one of the oldest and most important tasks in computer vision and photogrammetry. It is well-known that there are closed-form solutions for three-point problem (P3P problem) [82, 83], or generally PnP problem for any n points [84–86]. In addition, the PnP problem becomes the standard direct linear transformation method (DLT) when $n > 5$ [87, 88]. Wang and Jepson [89] solved the absolute orientation problem by forming linear combinations of constraint equations in such a way as to first eliminate the unknown rotation and then eliminate the unknown translation. Similar work was done by Fiore [90] where orthogonal decompositions were used to recover the point depths in the camera reference frame firstly. Wang [91, 92] proposed a solution procedure in which a calibration equation separating the rotational from translation parameters was involved. In this method, the rotation parameters were obtained first by solving a set of nonlinear equations, then the translation parameters were calculated analytically. There are some researchers who use lines [93] and orthogonal corners [94, 95] to provide a closed-form solution for the problem. On the other hand, bundle adjustment is an efficient way to achieve high accuracy in the estimation of the extrinsic parameters [96]. As a result, iterative pose estimation methods are also studied extensively as in [97–99]. For example, Taylor and Kriegman [97] minimized a nonlinear objective function with respect to camera rotation matrix, translation vector and 3D lines parameters. This method provides a robust solution to the high dimensional non-linear estimation problem. Lu et al. [100] formulated the problem by minimizing an object-space collinearity error and derived an orthogonal iterative algorithm for computing the pose and then the translation parameters.

Fundamental-Matrix Based Methods The fundamental matrix encapsulates the epipolar geometry of the imaging configuration. If we have calibrated the cameras and use the calibrated coordinate system, the fundamental matrix specializes to an essential matrix. It a 3×3 singular matrix and can be determined given eight point matches. Then the extrinsic parameters as well as the structure of a scene can be recovered. This is known as the classical eight-point algorithm [103, 104]. Using Singular Value Decomposition (SVD), Faugeras and Maybank [105] and Hartley [106] proposed a technique for factoring the fundamental matrix into a product of an orthogonal and a skew-symmetric matrix, which resulted in the solutions for the pose parameters. Kanatani and Matsunaga [107] addressed this problem based on an extensive study of the properties in the essential matrix. Ueshiba and Tomita [108] discussed three different cases to decompose the fundamental matrix. Some researchers studied the possibility to approach this problem with less than eight point matches, e.g. Triggs [109] and Nister [110, 111]. The point configurations, especially when less points are used, may be degenerate such that the algorithm fails to produce a unique solution. Philip [112, 113] investigated the 5-, 6-, 7-, and 8-point algorithms and suggested a way for handling the critical configurations.

The prominent advantage of the eight-point algorithm is that it is linear, hence fast and easy to implement. However, it is excessively sensitive to noise in the specification of the matched points which makes the estimation of the fundamental matrix critical. This is mainly due to two sources: the poor condition number of the coefficient matrix and the measurement error amplification. In 1997, Hartley [114] suggested a normalized eight-point algorithm to improve the condition number of the coefficient matrix. Recently, Wu et al. [115] found that the second drawback is mainly due to the product operation in the coefficient matrix, thus introduced a factorized eight-point algorithm and carried out a large amount of experiments to demonstrate its outstanding performance. Using the two eigenvectors corresponding to the two smallest eigenvalues achieved by the orthogonal least-squares technique, Zhong et al. [116] constructed a 3×3 generalized eigenvalue problem for estimating the fundamental matrix. Besides, there are two distinguished review articles regarding estimation of the fundamental matrix [117, 118].

Planar Homography-based Methods Planar surfaces are encountered frequently in some robotic tasks, e.g. in navigation of a mobile robot along ground plane and wall climbing robot for cleaning, inspection and maintenance of buildings. The eight-point algorithm will fail or give poor performance in the planar or near planar environments since they require a pair of images from the three-dimensional scene. Therefore, methods using only planar information need to be explored. Hay [119] was the first to report the observation that two planar surfaces undergoing different motions could give rise to the same image motion. Tsai and Huang [120] used the correspondences of at least four image points to determine the two interpretations of planar surfaces undergoing large motions, where a sixth-order polynomial of one variable was involved. A year late, Tsai et al. [121] approached the same problem by computing the singular value decomposition of a 3×3 matrix containing eight "pure parameters". Longuet-Higgins [122, 123] showed that three dimensional interpretations were obtainable by diagonalizing the 3×3 matrix, where the relative pose of the system and normal vector of the planar surface could be achieved simultaneously by a second-order polynomial. Given the planar Homography, Zhang and Hanson [124] proposed a method for this problem from a case by case analysis of different geometric situations, where as many as six cases were considered. According to the property of the planar Homography, Ma and Soatto et al. [125] decomposed it directly to the solution for the rotation matrix and translation vector. Habed and Boufama [126] formulated the self-calibration problem by solving the bivariate polynomial equations, where the points at infinity are parameterized in terms of the real eigenvalue of the Homography of the plane at infinity. In Sturm's work [127], beside a method for the basic one-view one-plane case, a factorization-based method for the multi-view multi-plane case possibly with little overlap was presented. Chum et al. [128] and Bartoli [129] independently investigated the relationships between the planar Homography and fundamental matrix and obtained some nice results regarding the consistency in them.

For the projector-camera system, Raij and Pollefeys [130] discussed a pose estimation technique for planar scenes to reconstruct the relative pose of a calibration camera, the projectors and the plane they projected on by treating the projectors as virtual cameras. When the coordinate system was chosen in some manner, Okatani and Deguchi [131, 132] showed that the screen-camera Homography as well as the poses of the projectors could be uniquely determined from only the images projected by the projectors and captured by the camera. Li and Chen [133] proposed a self-recalibration technique for a 2-DOF active vision system, where some constraints on the parameters of the system were explored, such as geometric constraints derived from similar triangular and focus constraints from best-focused location. Later, Chen and Li [134] gave another approach for 6-DOF active vision system by investigating the relationship between the camera's sensor plane and the projector's source plane. Another interesting technique is from Xie et al. [135], where a single-ball-target was used for the extrinsic calibration of their structured light system.

1.2.2 *Uncalibrated 3D Reconstruction*

1.2.2.1 Factorization-based Method

Here, the uncalibrated reconstruction means that the intrinsic parameters of the camera are unknown during the image acquisition process. Without any a priori knowledge, Faugeras [3] showed that at most a projective reconstruction could be obtained given a set of point correspondences. In his method, the epipolar geometry was estimated from the corresponding points on the images and the two projection matrices were computed but only up to a projectivity. Since then, a lot of researchers have studied the projective reconstruction from 2D images [4–6]. In 1995, Shashua [7, 8] investigated the multiple view geometry for 3D reconstruction in a very systematic and algebraic way. Especially, when working with three views or four views, the concepts of trifocal tensor or quadrifocal tensor which encapsulates the geometric relationship among three or four views are used [9].

When a large number of images, e.g. an image sequence, are available, the factorization-based method is preferred since it allows one to retrieve both shape and motion simultaneously from image measurements. Tomasi and Kanade [10] first proposed the factorization algorithm to achieve the maximum likelihood affine reconstruction with affine camera under the assumption of isotropic, mean-zero Gaussian noise, as pointed out by Reid and Murray [11]. In this sense, it is an optimal method if all the feature points are visible in each view. After that, the point based factorization method has been extended to lines and planes [12, 13, 136]. However, there are two critical issues in the factorization-based method that should be properly addressed.

One is the **projective scale** since the factorization is possible only when the image points are correctly scaled. Sturm and Triggs [14] proposed a practical method for the recovery of these scales using fundamental matrices and epipoles estimated from the image data. Some researchers also attacked this issue with iterative techniques [15] and bundle adjustment [16]. For example, Hung and Tang [16] proposed a factorization-based method that integrates the search of initial point and the projective bundle adjustment into a single algorithm. This approach minimizes the 2D reprojection error by solving a sequence of relaxed problems.

The other is related to **missing and outlier points**. Inevitably, some points will become occluded or out of field of view when changing the camera's viewpoint largely. Therefore, an efficient algorithm should be able to deal with this problem. Anas et al. [17] employed weighting factors to reduce the influence of outliers, where each weighting factor was set to be inversely proportional to the computed residual of each image feature point. There are also many articles that deal with this issue using different techniques, such as [18, 19].

1.2.2.2 Stratification-based Method

Due to the serious distortions, the application of projective reconstruction is limited. To remove the distortions, stratification-based method can be employed which defines a natural hierarchy of geometric structures: projective, affine and Euclidean structure. The challenging problem is to recover a transformation matrix $\tilde{\boldsymbol{T}}$ that upgrades the projective reconstruction to affine and then Euclidean reconstruction. Hartley [20] pioneered this kind of technique using a constant but uncalibrated camera. In this method, a projective reconstruction is firstly obtained from Levenberg-Marquardt minimization algorithm. To upgrade to the Euclidean structure, eight unknowns (three for affine and five for Euclidean) need to be recovered. Then three images are employed since each image excluding the first, gives six constraints. Heyden and Astrom [21] arrived at similar results in their work where the Kruppa constraints were derived. A 3×3 symmetric matrix equation is constructed which gives five equations on the eight parameters (the Homogeneity reduces the number of equations with one). With the assumption of constant intrinsic parameters, Pollefeys et al. [22] made use of modulus constraint to update the projective reconstruction to an affine one. In this method, a fourth order polynomial equation is involved for each image except the first, so a finite number of solutions can be found for four images.

In practice, the structured scene provides a lot of important information that can be incorporated with the stratified techniques [23–26]. In the work of Liebowitz and Zisserman [27], the image projections of parallel and orthogonal scene lines are detected and used for estimating the vanishing points as constraints in camera auto-calibration. Here, the fundamental matrix is involved, which in return limits the method to two images. A scheme is described in Heyden and Huynh [28] for the structure from motion problem, where constraints of orthogonal scene planes are imposed to recover the absolute quadric to upgrade the projective reconstruction to a Euclidean one. In this method, as many as images can be incorporated and a bundle adjustment algorithm is suggested to give a statistically optimal result. In the stratification-based method, perhaps the most difficult step is to obtain the affine reconstruction, or equivalently to locate the plane at infinity in the projective coordinate frame. Hartley et al. [29] tried to overcome this difficulty by imposing cheirality constraints to limit the search for the plane at infinity to a 3D cubic region of parameter space. Then an effective way of dense search over this cubic was carried out and a global minimum could be found robustly and rapidly.

1.2.2.3 Using Structured Light System

In the work of structured light systems, Scharstein [62] and Zhang et al. [63, 64] independently extended the traditional binocular stereo techniques for shape recovery from dynamic scenes. Their difference lies in that the former tries to decode the projected light patterns to yield pixel addresses in the projector while the latter simply sums up the error measures of all frames to directly compute stereo disparities. Recently, a unifying framework for the spacetime stereo was proposed in [65]. With a cubic frame, Chu et al. use cross-ratios and vanishing points for recovering unified world coordinates [66]. In this method, the intrinsic and extrinsic parameters of the camera and the position of the projector are not required.

On the other hand, by defining an image-to-world transformation between a 3D point on the stripe light plane and its corresponding pixel in the image, some novel methods have proposed for calibrating the structured light systems. To our knowledge, the earliest method for determining the image-to-world transformation was proposed by Chen and Kak [67] in 1987. They introduced a line-to-point method to calibrate their vision system, with a laser range finder mounted at the wrist of a robot arm and a flat trapezoidal object was put in the work area permanently. The object was shaped in such a manner that no two edges of the top-surface were parallel to each other. The end coordinates of the top edges of these objects were known to the robot or the equations that defined the lines corresponding to these edges were known. Then the robot moved the scanner to project a light plane onto the object. This generated a segment of the light stripe on the object whose two end points should lie on the two calibration lines respectively. Since each world line provides two equations with respective to the image-to-world transformation matrix, at least 6 world lines are required to calibrate the scanner system in their work.

Later, Reid [68] extended the concept to plane-to-point method. The laser range finder was mounted on the free-ranging AGV whose position relative to the world coordinate frame was known. With an appropriate choice of the stripe coordinate frame, the first row of the image to stripe transformation matrix was simplified to zero, leaving only 8 unknown parameters to be solved. The vehicle was driven to a known location where there existed an arrangement of orthogonal planes (A box with known size is used in that work) whose equations expressed in world coordinates had been pre-measured accurately. Known the vehicle position relative to the world coordinate frame and the sensor position relative to the vehicle coordinate frame, and a set of world plane to image point correspondences, an over-constrained set of linear equations were established to solve the required eight unknown parameters.

Recently, Huynh et al. [69] proposed a new point-to-point calibration method using four known non-coplanar sets of three collinear world points in the scene. In practice, the problem associated with calibrating a structured light system is that the known world points on the calibration object do not fall onto the light plane illuminated from the projector. So the invariance of cross ratio was employed to overcome this problem. Once four or more world points on each light plane were computed using cross ratio together with their corresponding image points, the image-to-world transformation matrix could be determined.

1.3 Scope

1.3.1 System Calibration

The camera calibration is the process of computing the intrinsic parameters (such as focal lengths, scale factors and principal point, etc) and the extrinsic parameters (position and orientation relative to a reference system) of the camera. Traditional static calibration methods tend to work in a rather cumbersome way: a calibration target with known dimensions, such as a movable checkerboard is presented to the system, and the positions of features on the calibration target are measured in the image. Whenever the system is adjusted or moved, it must be recalibrated again. Such calibration procedures are tedious and usually take considerable time and labor. On the other hand, the dynamic calibration algorithm which requires no special calibration targets or devices, has many distinct advantages. Firstly, it provides a fast and convenient mechanism. Although the system needs to be moved here and there, it can still be used as soon as it is placed, which is termed as "place-and-use". Secondly, when the size or distance of the object and the scene change, the camera's relative position and orientation also need be changed so that the camera can detect just the illuminated part of the scene. With a dynamic calibration method, the system is made ready for "place-and-use" without manual calibration. Lastly, in case of some hazardous places or when dealing with historical video data, instead of calibrating manually, the vision system must be calibrated dynamically without the operator's interferences.

1.3.2 Plane-based Homography

Planar structures are encountered frequently in our surroundings either indoors or outdoors, e.g. desks in the office, road surface, walls and ceilings of a building etc. According to the projective geometry, points on one plane are mapped to points on another plane by a plane-based Homography, also known as a planar projective transformation [137]. It is a bijective mapping induced by the bundle of rays centered in the camera centre. The Homographic matrix is a function about the imaging geometry (e.g. intrinsic and extrinsic parameters of the camera) as well as the scene plane, thus encapsulating almost all the information required in a vision system. So in our opinion, it is never too excessive to put emphasis on it although a great deal of work has been done on the planar Homography. In this book, we will use the planar structure as an important cue for dynamic calibration and 3D reconstruction via the plane-based Homography.

1.3.3 Structured Light System

The passive vision system usually consists of one or more CCD cameras. By passive, we mean that no energy is emitted for the sensing purpose. Hence the system receives the reflected light from its surroundings passively and the images are the only input data. On the whole, the passive system works in a similar way as human visual system does and its equipment is simple and low-cost. Another advantage is that the properties of the object's surface can be analyzed directly since it is based on ambient light reflection off the object surfaces. However, this type of techniques suffers from some nontrivial difficulties. Firstly, since the passive systems are mainly developed by utilizing the clues from the scene or illumination, e.g. texture or shade, the unknown scene becomes a source of the uncertainties. As a result, the major disadvantage is its low accuracy and speed. Secondly, to extract enough features and establish point correspondences from image sequence is a time-consuming and difficult task in many cases. Usually, there exists an awkward dilemma between the disparity and the overlapping area. With a large overlapping area between the images, it is easy to extract corresponding features. But the algorithm for 3D reconstruction is sensitive to noise or even ill-conditioned since the images have small disparity. On the other hand, the algorithm becomes more and more robust as the disparity is enlarged. But it is difficult to extract the corresponding features due to occlusion, or out of scope of viewing field, etc.

While the passive vision system suffers from many difficulty problems, the active vision system especially structured light system, is designed to overcome or alleviate these problems to a certain degree. In this kind of system, one of the cameras in passive system is replaced by an external projecting device (e.g. laser or LCD/DLP projector). The scene is illuminated by emitting light patterns from the device and detected by the remaining cameras. Compared with the passive approach, the active vision systems are in general more accurate and reliable.

In this book, we will introduce a structured light system that consists of a CCD camera and a DLP projector. In such a system, the features we use are regularly created by the light patterns. It is easy to extract these features since we know a prior what kind of features they are. And the corresponding problem is ready to be solved since the light patterns are appropriately encoded with a pattern coding strategy.

1.3.4 Omni-Directional Vision System

Both the traditional vision system and the structured light system often suffer from limited field of view. There are many different approaches to enlarge the viewing angle of a vision system. Among those, the cooperation of a specially-shaped mirror with a conventional lens or fisheye lens camera is the most effective way. Such system is usually named omni-directional vision system or catadioptric vision system. In this book, we will introduce the configurations of three catadioptric vision systems, together with their imaging models and parameter calibration via various cues from their surroundings. After that, the needed 3D information can be reconstructed from their images.

1.4 Objectives

The main objective of this book is the design, implementation and evaluation of various vision systems, which mainly includes the following four tasks:

1. Setting up a structured light system and applying it to 3D reconstruction;
   With a DLP projector, this book talks about a unique color-encoded method which

provides a fast way to identify the coordinates on the projector plane by only local analysis. Here, the projector is modeled in two different ways: One is considered as a virtual camera and the other is as a collection of light planes. Within this framework, it is possible for the vision system to recover the 3D surface by taking only a single image so that the vision tasks can be performed at the frame rate speed.

2. Constructing a catadioptric vision system and investigating its imaging model;
   To enlarge the field of view, different configurations in the catadioptric vision systems are discussed, including panoramic stereoscopic system that consists of a convex mirror and a CCD camera with fisheye lens, parabolic camera system that consists of a parabolic mirror and a CCD camera with orthographic projection, and hyperbolic camera system that consists of a hyperbolic mirror and a perspective camera. In these systems, the central projection with a single effective viewpoint is ensured considering the characteristics of the mirrors.
3. Studying various methodologies for dynamic calibration of those systems by exploring the plane-based Homography and fundamental matrix;
   Since our system is designed to be applicable for static as well as dynamic scenes with moving or deformable objects, the time cost is critical. So the developed methods should be linear and provide an analytic solution to avoid the time-consuming nonlinear optimization. The geometric properties of scene structures and imaging model of the systems are extensively analyzed to provide necessary constraints on the system parameters.
4. Carrying out both numerical simulations and real data experiments to validate the constructed systems.
   In general, the accuracy of 3D reconstruction depends heavily on that of the calibration. In the presence of noise, any small errors on the calibration results will propagate to the 3D reconstruction. Therefore, how to improve the accuracy of the calibration and evaluate its errors is particularly important in a system that calibration is frequently required. Some results of the numerical simulations and real data experiments are presented with our vision systems.

1.5 Book Structures

The remainder part of the book is organized as follows:

Chapter 2 introduces mathematical models for the components in a structured light system and catadioptric system. The coding and decoding strategies for the light pattern are presented. Some preliminaries such as plane-based Homography, fundamental matrix and cross ratio, are presented.

In Chap. 3, we firstly discuss basic theory for camera calibration based on the image of absolute conic. The task is to estimate intrinsic parameters of the camera. Three planar patterns which respectively contain equiangular polygon, intersectant circles and concentric circles, are designed when implementing the theory.

Chapter 4 deals with the relative pose problem in the structured light system with the intrinsic parameters calibrated. By assuming there is an arbitrary plane, the image-to-image Homography is extensively explored to provide an analytic solution. The error sensitivity of the estimated pose parameters with respect to the noise in the image points is also investigated.

In Chap. 5, the projector in the structured light system is deemed as a collection of light planes. We talk about the image-to-world transformation and its use in obtaining 3D structure of the scene. Based on the two-known-plane assumption, a method for dynamically determining the transformation matrix is discussed. Besides, the computational complexity is analyzed and compared with the traditional method. In case of one-known-plane assumption, we give a scheme for calibrating the focal lengths and extrinsic parameters of the camera using the plane-to-image Homography, and then the image-to-world transformation.

In Chap. 6, the configuration and mathematical models of the catadioptric vision system are investigated. Approaches for obtaining a central projection process in the system are discussed. Algorithms for calibrating the intrinsic as well as extrinsic parameters of the system are investigated.

Finally, some conclusions and suggestions for future work are presented in Chap. 7.

Chapter 2
System Description

Most vision systems are currently task-oriented, which means that different systems are configured for different vision tasks in various environments. Anyway, a good vision system should make full use of all the merits of each component within the system. The main objective of this chapter is to provide a brief introduction of various components involved in the concerned vision system, i.e. structured light system and omni-directional vision system. And their characteristics are analyzed at the same time. Firstly, we introduce the mathematical models for the mirror, the camera and the projector, respectively. Then the coding and decoding strategies for the light pattern of the projector are discussed. Finally, we present some preliminaries that will be used in the succeeding chapters.

2.1 System Introduction

In this book, we consider two different kinds of vision systems. One is the structured light system which can actively illuminate its environment with a predefined light pattern. The distinct advantage of such system is that desired features can be created via the design of the pattern. The other is omni-directional vision system, in which a specially-shaped mirror or lens is generally involved. Hence, such system possesses a much larger field of view than the traditional system.

2.1.1 Structured Light System

In general, the classic stereo vision system is configured with one or more cameras (Fig. 2.1). Before a vision task starts, such as 3D reconstruction, feature points should be tracked among the images or image sequence. Due to occlusions or changing in element characteristics between different images, some of those features may be lost or false matching. Furthermore, it is very difficult to obtain dense matches in some cases. This is named as the correspondence problem.

B. Zhang, Y. F. Li, *Automatic Calibration and Reconstruction for Active Vision Systems*,
Intelligent Systems, Control and Automation: Science and Engineering,
DOI 10.1007/978-94-007-2654-3_2,

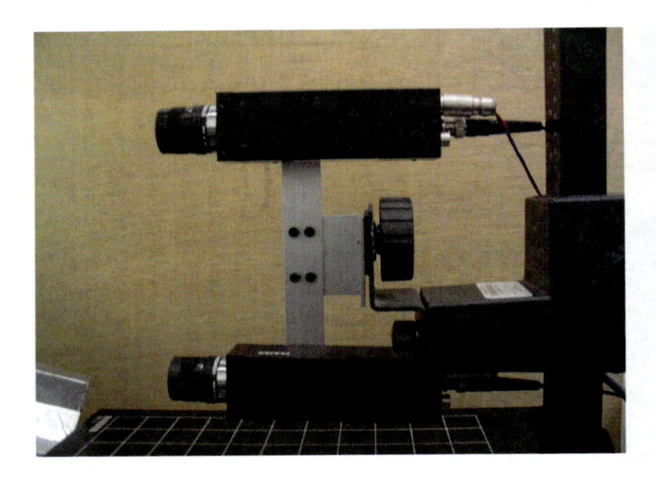

Fig. 2.1 A traditional stereo vision system

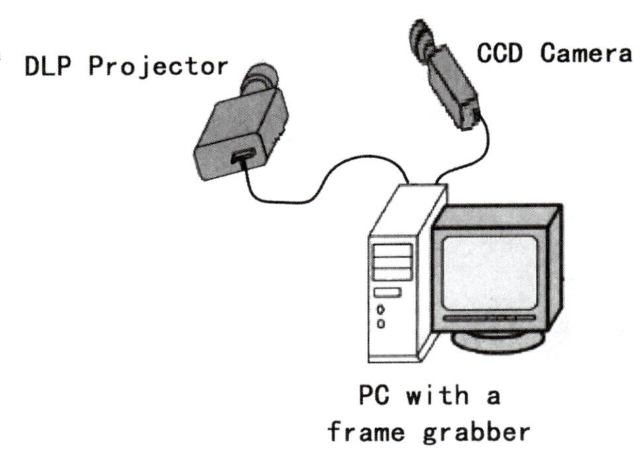

Fig. 2.2 Configuration of a typical structured light system

By replacing one of the cameras in the traditional stereo system with an illumination device, we can obtain a new system usually named the structured light system. The distinct advantage lies in that it provides an elegant solution for the correspondence problem. In practice, there are many alternative illumination devices, such as a laser (Reid [68]), a desk lamp (Bouguet [73]) or a DLP projector (Okatani [132]), which can be used to create desired feature points with a predefined light pattern in the surroundings. Figure 2.2 shows a sketch of a typical structured light system consisting of a CCD camera, a DLP projector and a personal computer with the frame grabber.

When working, the projector which is controlled by a light pattern, projects a bundle of light spots, light planes or light grids into the scene, which in return are reflected by the scene's surface and sensed by the camera as an image (Fig. 2.3). In such scenario, the feature correspondences are easily solved since we already know where they come from according to the design of the light pattern. And dense feature

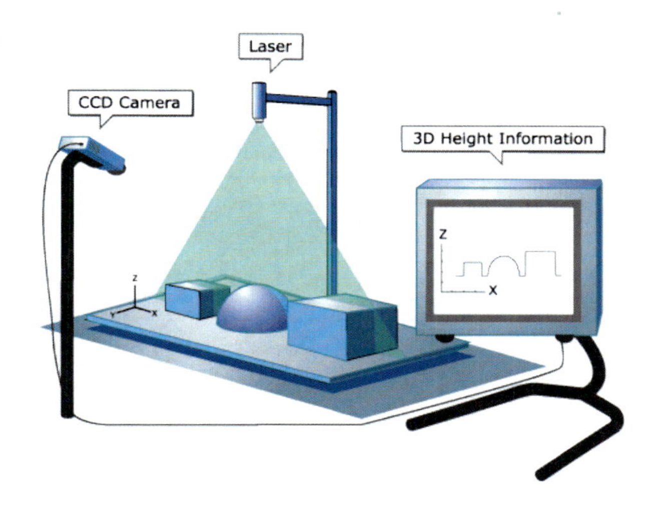

Fig. 2.3 Profile of a system. (From http://www.stockeryale.com/i/lasers/structured_light.htm)

points can be extracted on each light stripe in the image. When the vision system is calibrated, 3D modeling of the scene or concerned objects can be carried out with classical triangulation method.

2.1.2 Omni-Directional Vision System

The conventional structured light systems generally have limited fields of view, which make them restrictive for certain applications in computational vision. This shortcoming can be compensated through an omni-directional vision system, which combines the refraction and reflection of light rays, usually via lenses (dioptrics) and specially curved mirrors (catoptrics). The angle of view in such system is larger than 180° or even $360 \times 360^{\circ}$, enabling to capture the entire spherical field of view. In practice, there are various configurations. Figure 2.4 shows two typical systems and their images from [178].

When working, the light rays are firstly reflected by the mirror according to the law of reflection. Then the reflected rays are sensed by the camera into an image. Once the system has been calibrated, 3D reconstruction can be similarly implemented by classical triangulation algorithm.

2.2 Component Modeling

2.2.1 Convex Mirror

Many kinds of mirrors, such as planar mirror, Pyramidal multi-faceted mirror, conical and spherical mirror, can be used in an omni-directional vision system to obtain a larger field of view. A plane mirror is simply a mirror with a flat surface while the rest

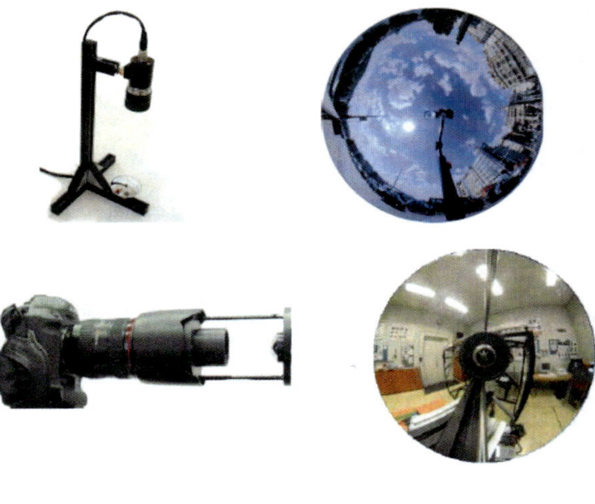

Fig. 2.4 Two configuration of the catadioptric systems and their images

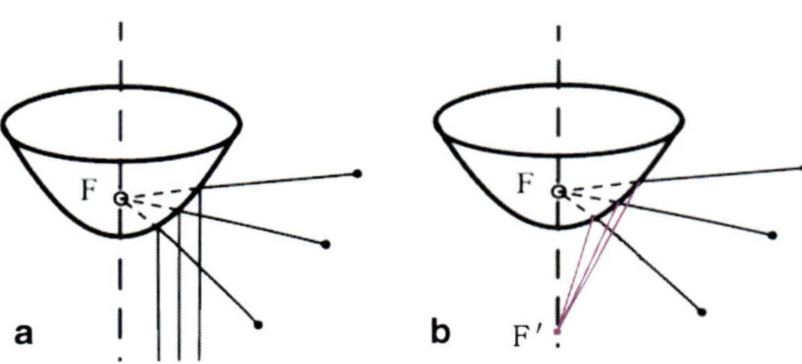

Fig. 2.5 **a** The parabolic mirror and **b** hyperbolic mirror

have much complex-shaped surface. They all follow the law of reflection, which says that the direction of the incoming light (incident ray), and the direction of outgoing light (reflected ray) make the same angle with respect to the surface normal. In this book, we mainly employ two types of mirrors, i.e. the parabolic mirror and the hyperbolic mirror.

The parabolic mirror has a particular kind of three-dimensional surface. In the simplest case, it is generated by the revolution of a parabola along its axis of symmetry as illustrated in Fig. 2.5a. Here, the dashed vertical line represents the symmetry axis and $\boldsymbol{F}$ is the focal point. According to its mathematical model, the incident rays will pass through the focal point while their reflected rays will be parallel with the symmetric axis. The parabolic-mirror-camera system, the parabolic camera system for short, is a vision system that incorporates the parabolic mirror with orthogonal projection camera. Certainly, all the light rays round the parabolic mirror can be reflected by its surface and sensed by the camera into an image. Hence, the potential horizontal field of view is 360°.

In mathematics, a hyperboloid is a quadratic surface of revolution which may have one or two sheets. And the symmetry axis passes through its two foci, denoted by $\boldsymbol{F}$ and $\boldsymbol{F'}$ (referring to Fig. 2.5b). The hyperbolic mirror follows the shape of one sheet of a hyperboloid, in which the trajectory of a light ray is: an incoming ray

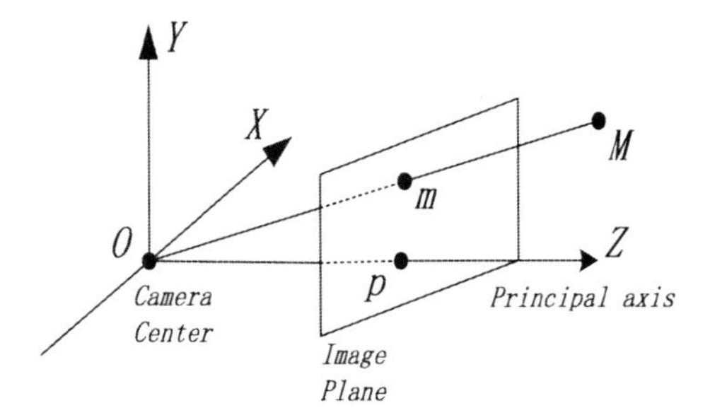

Fig. 2.6 Illustration of a pinhole camera model

passing through the first focal point, say $\boldsymbol{F}$, should be reflected such that the outgoing ray will converge at the second focal points, say $\boldsymbol{F}'$. The hyperbolic-mirror-camera system, the hyperbolic camera system for short, is a vision system that combines a hyperbolic mirror and a pinhole camera that is placed at the point $\boldsymbol{F}'$. From the ray trajectory, we can see that this type of system has one effective viewpoint and the horizontal field of view is 360°.

2.2.2 Camera Model

A camera is a mapping from the 3D world to 2D image plane. In this book, the perspective camera model is used, which corresponds to an ideal pinhole model. Figure 2.6 illustrates the geometric process for image formation in a pinhole camera. Here, the centre of projection, generally called the camera centre, is placed at the origin of coordinate system. The line from the camera centre perpendicular to the image plane is called the principal axis, while their intersection is called the principal point. Mathematically, the camera model can be represented by the following 3 × 3 nonsingular camera matrix ([150], see Chap. 5):

$$\boldsymbol{K}_c = \begin{bmatrix} f_u & s & u_0 \\ 0 & f_v & v_0 \\ 0 & 0 & 1 \end{bmatrix} \tag{2.1}$$

where: f_u and f_v represent the focal lengths of the camera in terms of pixel dimensions along U-axis and V-axis respectively, and $(u_0 \quad v_0)^{\mathrm{T}}$ is the principal point, s is a skew factor of the camera, representing cosine value of the subtending angle between U-axis and V-axis. Since there are five parameters in this model, it is referred to as a linear five-parameter camera. The ratio of f_u and f_v is often called the aspect ratio of the camera. With a modern camera, the skew can be treated as zero ($s = 0$) which means the pixels in the image can be assumed to be rectangular. In this case, the model is referred to as a four-parameter camera.

For a camera with fixed optics, these parameters are identical for all the images taken with the camera. For a camera which has zooming and focusing capabilities, the focal lengths can obviously change. However, as the principal point is the intersecting point of principal axis with the image, it can be assumed to be unchanged sometimes. It is reported that the assumptions are fulfilled to a sufficient extent in practice [151, 152]. Recently, Sturm et al. [153], Cao et al. [154] and Kanatani et al. [155] used these assumptions by permanently setting the principal point to (0, 0), the skew to zero and the aspect ratio to one in their work. Another interesting work is from Borghese et al. [156], where only the focal length is unknown and can be computed with cross ratio, based on zooming in and out a single 3D point.

So in our work, when dynamically calibrating the intrinsic parameters, we always assume that the focal lengths in both pixel dimensions are unknown and variable, i.e. the camera matrix can be simplified as

$$\boldsymbol{K}_c = \begin{bmatrix} f_u & 0 & 0 \\ 0 & f_v & 0 \\ 0 & 0 & 1 \end{bmatrix} \tag{2.2}$$

In general, the camera coordinate system does not coincide with the world coordinate system and their relationship can be described by a rotation matrix and a translation vector, denoted by $\boldsymbol{R}_c$ and $\boldsymbol{t}_c$ respectively. Let $\boldsymbol{M}$ be a 3D point in the world coordinate system and $\tilde{\boldsymbol{M}}$ its corresponding homogeneous representation. Under the pinhole camera model, its projection point $\tilde{\boldsymbol{m}}$ in the image is given by

$$\begin{aligned} \tilde{\boldsymbol{m}} &= \beta \boldsymbol{K}_c(\boldsymbol{R}_c\boldsymbol{M} + \boldsymbol{t}_c) \\ &= \beta \boldsymbol{K}_c[\boldsymbol{R}_c \quad \boldsymbol{t}_c]\tilde{\boldsymbol{M}} \end{aligned} \tag{2.3}$$

where β is a nonzero scale factor.

2.2.3 *Projector Model*

In general, a light projector can be treated as the dual of a camera, i.e. a projection device relating the 3D world and the 2D image. So the projector image formation can be approximated by a pin-hole projection model, similar to the pin-hole camera model. This means that the projector can be described by the following 3×3 matrix

$$\boldsymbol{K}_p = \begin{bmatrix} f'_u & s' & u'_0 \\ 0 & f'_v & v'_0 \\ 0 & 0 & 1 \end{bmatrix} \tag{2.4}$$

with f'_u, f'_v, s', u'_0 and v'_0 defined similar to those in (2.1).

However, the assumption about a simplified camera model, that the principal point is close to the image center, is not valid for a projector since most projectors use an off-axis projection. For example, when they are set on a table or mounted upside-down on a ceiling, the image is projected through the upper or

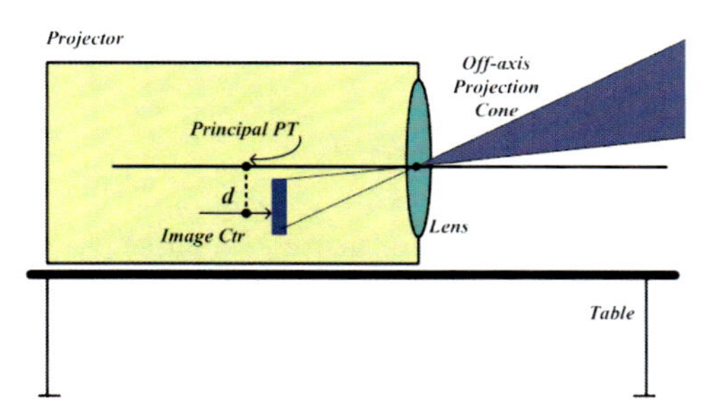

Fig. 2.7 The optical characteristic of a projector

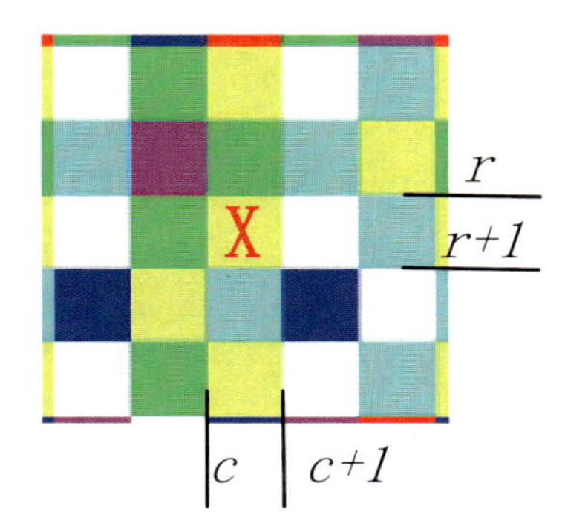

Fig. 2.8 Identification of light plane's index

the lower half of the lens, respectively [157]. As a result, the principal point is vertically shifted or even slightly outside the image. Therefore, we always assume that the projector has fixed optics, which means its intrinsic parameters are kept constant.

In the structured light system, the light pattern (to be discussed later) for the projector is designed as a color-encoded grid, which contains two sets of parallel lines perpendicular to each other. The trace of each line forms a plane in 3D space, named as a stripe light plane. So another straightforward model for the projector is to describe it by two sets of stripe light planes. To identify the index of each light plane, the coordinates of the grid associate with that plane are obtained first. For example, if the coordinates of the grid 'X' are (r, c), the index of the upper horizontal plane associated with the grid is set to be r and that of the left vertical plane is c (Fig. 2.8 shows a portion of the light pattern). Then the indexes for other consecutive planes can be retrieved by simple plus and minus operations.

In summary, two different models for the DLP projector are established. The first model is used in Chap. 4 by treating the system as an active stereoscopic vision, while the second model is used in Chap. 5 by treating the system as a collection of light planes and a camera.

2.3 Pattern Coding Strategy

2.3.1 Introduction

Studies on the active vision techniques for computation of range data can be traced back to the early 1970s from the work of Shirai and his collaborators [138, 139]. The light pattern used in the vision system ranges from a single point and a single line segment to a more complicated pattern, such as a set of parallel stripes and orthogonal grids, etc. A review work on the recent development in coded structured light techniques can be found in [149]. Broadly speaking, they can be classified into two categories: static coding and dynamic coding techniques.

The static coding methods include binary codes [140] and Gray codes [141, 142], etc, where a sequence of patterns are used to generate the codeword for each pixel. This kind of techniques is often called time-multiplexing methods because the bits of the codeword are multiplexed in time. The advantage is that the pattern is simple and the resolution and accuracy of the reconstructed data can be very high. However, it is limited to static scenes with motionless objects, and hence termed as static coding.

On the other hand, the dynamic coding methods are developed based on De Bruijn sequences [143, 144] and M-arrays [145, 146], etc, in which the light pattern is encoded into a single shot. In general, spatial neighborhood strategy is employed so that the light pattern is divided into a certain number of regions, in which some information generates a different codeword. The major advantage is that such strategy permits a static as well as dynamic scene with moving or deformable objects.

In our work, we require that a single image should be sufficient for the calibration and 3D reconstruction, which implies that the light pattern for the DLP projector should be encoded into a single shot. Here, a dynamic coding method from Griffin [147, 148] is adopted. The algorithms can be summarized as follows.

2.3.2 Color-Encoded Light Pattern

Let $\omega = \{1, 2, \ldots, \gamma\}$ be a set of color primitives (for example, 1 = red, 2 = green, 3 = blue, etc). Given these color primitives, a one-dimensional string V_{hp} is constructed such that each triplet of adjacent colors is distinct from every other triplet. This string is considered as the first row in the light pattern and its size is $\gamma^3 + 2$. Similarly, another string V_{vp} is constructed such that each pair of adjacent colors is distinct from every other pair and the size is $\gamma^2 + 1$. For the other rows in the pattern, modulo operation is iteratively performed with the preceding row and each element of the second string. Then we have a matrix for the light pattern whose size is $(\gamma^2 + 2) \times (\gamma^3 + 2)$. For example, if $\omega = \{1, 2, 3\}$ is taken, the following two vectors are obtained:

$$V_{hp} = (331 \quad 321 \quad 311 \quad 231 \quad 221 \quad 211 \quad 133 \quad 232 \quad 223 \quad 33)$$
$$V_{vp} = (31 \quad 21 \quad 13 \quad 22 \quad 33)$$

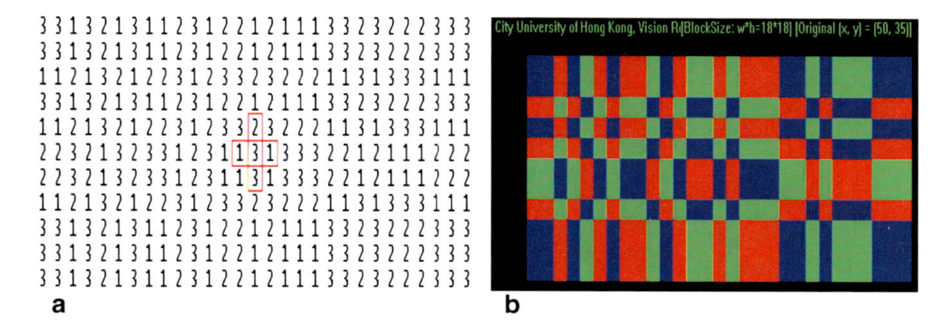

Fig. 2.9 Example of an encoded matrix (Three color primitives). **a** The pattern matrix. **b** A screen shot of the color-encoded light pattern

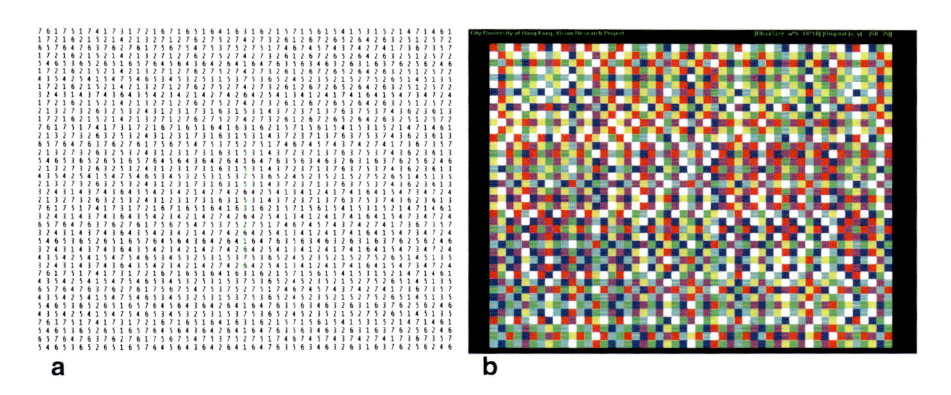

Fig. 2.10 Example of an encoded matrix (Seven color primitives). **a** The pattern matrix. **b** A screen shot of the color-encoded light pattern

Then, the first row of the pattern is $p_{0i} = V_{hpi}$. The rest of the matrix elements are calculated using

$$p_{ij} = (p_{(i-1)j} + V_{vpj}) \text{mod } 3) + 1 \tag{2.5}$$

Finally, a matrix with size 11 × 29 is obtained as shown in Fig. 2.9a, b gives a screen shot of the light pattern. Here, a codeword at location (i, j) is defined by the color primitive and its four neighbors (east, south, west and north). Griffin [147] had proved that all the codewords are unique in this pattern.

From this figure, we can see that some adjacent grids may have the same color and will not be distinguished from each other. So we use a codeword-based filtering procedure to discard those grids having the same color with their neighbors. In our system, we take $\omega = \{1, 2, \ldots, 7\}$ where seven different colors, i.e. red, green, blue; white, cyan, magenta; yellow, are used. According to the formula, we can have a matrix of 51 × 345. Among which, those grids who are distinct with their neighbors are selected. The final results are shown in Fig. 2.10. Here, 40 × 51 matrix is used.

Table 2.1 Look-up table for horizontal triplets

Triplet	Column index	Triplet	Column index	Triplet	Column index
(3,3,1)	2	(2,3,1)	11	(1,3,3)	20
(3,1,3)	3	(3,1,2)	12	(3,3,2)	21
(1,3,2)	4	(1,2,2)	13	(3,2,3)	22
(3,2,1)	5	(2,2,1)	14	(2,3,2)	23
(2,1,3)	6	(2,1,2)	15	(3,2,2)	24
(1,3,1)	7	(1,2,1)	16	(2,2,2)	25
(3,1,1)	8	(2,1,1)	17	(2,2,3)	26
(1,1,2)	9	(1,1,1)	18	(2,3,3)	27
(1,2,3)	10	(1,1,3)	19	(3,3,3)	28

Remark

1. It should be noted that rather than color dots used in Griffin [147], we use the color-encoded grid blocks. In our opinion, the grid blocks can be segmented more easily by edge detection. The encoded points are the intersection of these edges, so they can be found very accurately. When projecting dots, their mass centres must be located. When a dot appears only partially in the image or the surface of the object is somewhat bumpy, the mass centre will be incorrect. Obviously, our light pattern overcomes these drawbacks. Moreover, the grid techniques allow adjacent cross-points to be located by tracking the edges, but this is not applicable for the dot representation. These features not only decrease the complexity of image processing but also simplify the pattern decoding process.
2. There is a trade off between the resolution of the pattern, i.e. the number of color primitives used and the complexity of image processing. In general, the larger the number of color primitives used, the higher the resolution we will have and more details of the scene can be captured, but the more the noise sensitivity and hence the higher the complexity of image processing. Through our experiments, we find that a good balance can be obtained when seven color primitives are used. Anyway, an interpolation operation can provide an approximation of dense information when required.

2.3.3 Decoding the Light Pattern

Once the light pattern is projected on the concerning scene, an image is grabbed with the camera. The decoding problem involves finding the coordinates of the codewords extracted from the image. For simplicity, we take the three-color-primitive case in Fig. 2.9 as an example.

Before decoding, two tables are constructed. Table 2.1 consists of a list of each triplet in V_{hp} as well as the column position of the middle primitive of each triplet. Table 2.2 consists of a list of each pair in V_{vp} along with the row position of the first

Table 2.2 Look-up table for vertical pairs

Vertical pairs	(3,1)	(1,2)	(2,1)	(1,1)	(1,3)	(3,2)	(2,2)	(2,3)	(3,3)
Cumul. jumps	3	4	6	7	8	11	13	15	18
Row index	2	3	4	5	6	7	8	9	10

primitive of each pair and the cumulative jump. The cumulative jump value is defined as $c\delta_i = c\delta_{i-1} + \delta_i$, where δ_i is the first element value of the pair at position i.

After the preprocessing stage, the algorithm for decoding a codeword w_{ij} is presented as follows:

1. Determine the horizontal triplet of $w_{ij} : (h_1, h_2, h_3) = (w_{i,j-1}, w_{i,j}, w_{i,j+1})$;
2. Determine the vertical triplet of $w_{ij} : (v_1, v_2, v_3) = (w_{i-1,j}, w_{i,j}, w_{i+1,j})$;
3. Calculate the number of jumps (a, b) between the primitives:
   $a = (v_2 - v_1) \bmod \gamma$ and $b = (v_3 - v_2) \bmod \gamma$
   then identify the row index i from Table 2.2 which corresponds to (a, b) and get the cumulative jump $c\delta$ for this value;
4. Determine the alias (a_1, a_2, a_3) by:

$$b\delta = (c\delta) \bmod \gamma$$

$$a_1 = (h_1 - b\delta) \bmod \gamma$$

$$a_2 = (h_2 - b\delta) \bmod \gamma$$

$$a_3 = (h_3 - b\delta) \bmod \gamma$$

Then identify the column index j by locating the alias in Table 2.1.

We can obtain the coordinate (i, j) for the codeword w_{ij}.

Note that in the above algorithm, if $(k - l) \leq 0$ then $(k - l) \bmod \gamma = k - l + \gamma$.

For example, we consider the codeword (3, 1, 2, 1, 2) labeled by red line in Fig. 2.9. For this codeword, the vertical triplet is (2, 3, 3), so the number of jumps (a, b) is (1, 3). From Table 2.2, the pair (1,3) matches the row index $i = 6$ and the cumulative jumps $c\delta = 8$. Therefore, $b\delta = 2$. From the horizontal triplet (1, 3, 1), we get the vertical alias $(a_1, a_2, a_3) = (2, 1, 2)$. From Table 2.1, the column index is $j = 15$. Consequently, the coordinates for the codeword is (6, 15). From the light pattern matrix, we can verify that it is correct.

2.4 Some Preliminaries

2.4.1 Notations and Definitions

I. Notations $\boldsymbol{\Pi}, \boldsymbol{\Pi}_k, \ldots,$ represent the projective planes: The image plane of the CCD camera is denoted by $\boldsymbol{\Pi}$, and the k-th light stripe plane by $\boldsymbol{\Pi}_k$.

$\boldsymbol{F}_w, \boldsymbol{F}_{\Pi}, \boldsymbol{F}_{2\Pi k}, \boldsymbol{F}_{3\Pi k}, \ldots,$ represent the coordinate frames. The subscript "2" specifies a 2D coordinate frame, and "3" defines a 3D coordinate frame. We denote the world coordinate frame by $\boldsymbol{F}_w$, the image coordinate system by $\boldsymbol{F}_{\Pi}$, and the coordinate frame of the k-th stripe light plane by $\boldsymbol{F}_{2\Pi k}$ and $\boldsymbol{F}_{3\Pi k}$.

The scale factor is denoted by normal face symbols, e.g. a. A vector represents a column of real numbers denoted by lowercase boldface letter, such as $\boldsymbol{a}$. If not specified, we mean it a 3×1 vector. A matrix is an array of real numbers denoted by uppercase boldface letter, such as $\boldsymbol{A}$. If not specified, we mean it a 3×3 matrix. Superscript T in Greek style represents the transpose of a vector or matrix.

The boldface letters $\mathbf{0}$, $\boldsymbol{I}$, $\boldsymbol{R}_c$ and $\boldsymbol{t}_c$ always denotes the zero matrix, identity matrix, rotation matrix and translation vector, respectively, while the notations $\boldsymbol{M}$ and $\boldsymbol{m}$ denote the 3D and 2D points.

The symbol $[*]_\times$ represents skew symmetric matrix of a vector $*$. For example, if

$$\boldsymbol{t}_c = (t_1 \quad t_2 \quad t_3)^{\mathrm{T}} \quad \text{then} \quad [\boldsymbol{t}_c]_\times = \begin{bmatrix} 0 & -t_3 & t_2 \\ t_3 & 0 & -t_1 \\ -t_2 & t_1 & 0 \end{bmatrix}.$$

When necessary, further notation choices are described in each chapter.

II. Definitions In 2D Euclidean space, a 2-vector $\boldsymbol{m} = (x \quad y)^{\mathrm{T}}$ can be used to represent the inhomogeneous coordinates of a point $\boldsymbol{m}$. By adding a final coordinate with value 1, the 3-vector $\tilde{\boldsymbol{m}} = (x \quad y \quad 1)^{\mathrm{T}}$ becomes a homogeneous representation of that point. On the other hand, if the 3-vector $(x_1 \; x_2 \; x_3)^{\mathrm{T}}$ represents a 2D point, its inhomogeneous coordinates is given by $(\frac{x_1}{x_3} \quad \frac{x_2}{x_3})^{\mathrm{T}}$. In the homogeneous representation, if the final coordinate is zero or close to zero, e.g. $x_3 = 0$, the inhomogeneous coordinates will be infinity or close to infinity. Such point is called point at infinity. The line at infinity is such a line that only consists of the points at infinity. The 2D Euclidean space and the line at infinity make up a 2D projective space.

In the 2D projective space, the joint or cross product of two points gives a line in the same space. This line is called the line at infinity if the two points are both point at infinity. Dually, the intersection of any two lines provides a point. When the two lines are parallel to each other, this point is the point at infinity. It depends only on the direction of those lines, but not their positions. In vision community, the image of a line at infinity is named a vanishing line while the image of a point at infinity a vanishing point. Similar results can be obtained for 3D projective space which consists of 3D Euclidean space and the plane at infinity.

2.4.2 Cross Ratio

In definition, the cross ratio is a ratio of ratios of distances. Given four collinear points in 3D space, $\boldsymbol{A}_1$, $\boldsymbol{A}_2$, $\boldsymbol{A}_3$ and $\boldsymbol{A}_4$, the cross ratio is expressed as

$$(\boldsymbol{A}_1, \boldsymbol{A}_2; \boldsymbol{A}_3, \boldsymbol{A}_4) = \frac{\overline{\boldsymbol{A}_1\boldsymbol{A}_3}}{\overline{\boldsymbol{A}_2\boldsymbol{A}_3}} \cdot \frac{\overline{\boldsymbol{A}_2\boldsymbol{A}_4}}{\overline{\boldsymbol{A}_1\boldsymbol{A}_4}} \tag{2.6}$$

where $\overline{\boldsymbol{A}_1\boldsymbol{A}_3}$ denotes the length between $\boldsymbol{A}_1$ and $\boldsymbol{A}_3$, etc.

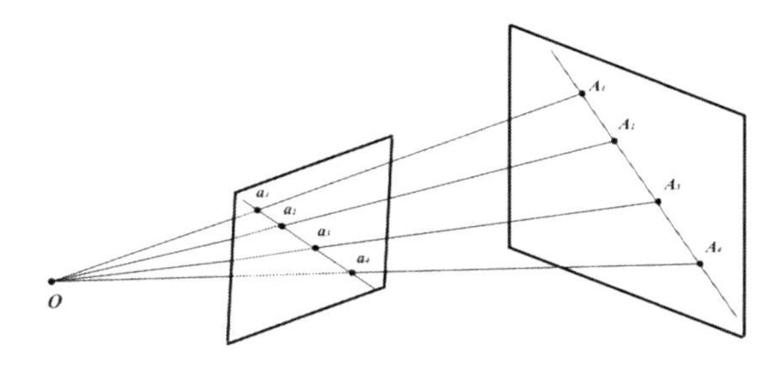

Fig. 2.11 Invariance of cross ratio of four collinear points under perspective projection

If their correspondent points on the projective plane (such as CCD camera image plane) are $\boldsymbol{a}_1$, $\boldsymbol{a}_2$, $\boldsymbol{a}_3$ and $\boldsymbol{a}_4$, according to the property of projective geometry, they are also collinear (Fig. 2.11). As cross ratio is invariant under projective transformation, we have the following equation:

$$(\boldsymbol{A}_1, \boldsymbol{A}_2; \boldsymbol{A}_3, \boldsymbol{A}_4) = (\boldsymbol{a}_1, \boldsymbol{a}_2; \boldsymbol{a}_3, \boldsymbol{a}_4) \tag{2.7}$$

The cross ratio is independent of the coordinate system established. Especially, if the points are parameterized by θ_{A_1}, θ_{A_2}, θ_{A_3} and θ_{A4}, then

$$(\boldsymbol{A}_1, \boldsymbol{A}_2; \boldsymbol{A}_3, \boldsymbol{A}_4) = \frac{\theta_{\boldsymbol{A}_1} - \theta_{\boldsymbol{A}_3}}{\theta_{\boldsymbol{A}_2} - \theta_{\boldsymbol{A}_3}} \cdot \frac{\vartheta_{\boldsymbol{A}_2} - \theta_{\boldsymbol{A}_4}}{\theta_{\boldsymbol{A}_3} - \theta_{\boldsymbol{A}_4}} \tag{2.8}$$

Here, the cross ratio is defined in terms of collinear points. This definition can be extended to the pencil of lines and pencil of planes. With the tool of cross ratio, we can calculate a 3D point through its image point coordinate. This is very useful when calibrating the light planes as in Chap. 5.

2.4.3 Plane-Based Homography

According to the projective geometry, the term plane-based Homography refers to the plane-to-plane transformation in the projective space, which maps a point on one plane to a point on the other. The Homography arises when a planar surface is imaged or two views of the planar surface are obtained. Figure 2.12 shows one of the cases, in which $\boldsymbol{M}$ represents a space point on the plane π, $\boldsymbol{m}_c$, and $\boldsymbol{m}_p$ denote its corresponding projections.

Algebraically, the Homography can be described by a 3×3 non-singular matrix, e.g. $\boldsymbol{H} = \begin{bmatrix} h_1 & h_2 & h_3 \\ h_4 & h_5 & h_6 \\ h_7 & h_8 & h_9 \end{bmatrix}$. Under the perspective projection, the corresponding points $\tilde{\boldsymbol{m}}_p$ and $\tilde{\boldsymbol{m}}_c$ are related by

$$\tilde{\boldsymbol{m}}_p = \lambda \boldsymbol{H} \tilde{\boldsymbol{m}}_c \tag{2.9}$$

where λ is a scale factor.

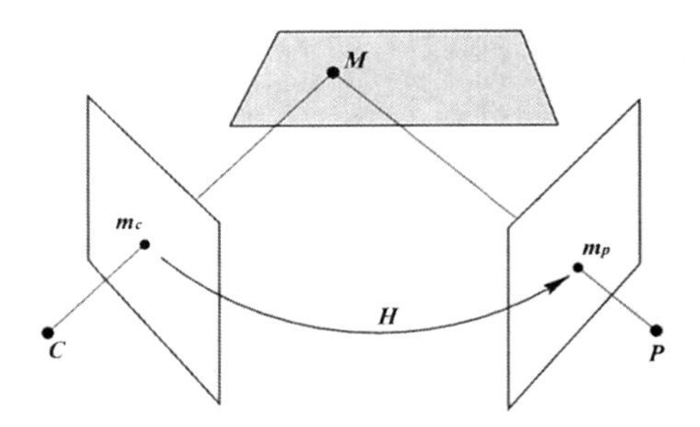

Fig. 2.12 An illustration for plane-based Homography

From (2.9), each pair of corresponding points provides two constraints on the Homography. Hence, four pairs are sufficient for the estimation. If more than four pairs are available, the Homography can be determined in the least squared sense. Let the i-th pair of points be $\tilde{\boldsymbol{m}}_{c,i} = [u_i \quad v_i \quad 1]^{\mathrm{T}}$ and $\tilde{\boldsymbol{m}}_{p,i} = [u'_i \quad v'_i \quad 1]^{\mathrm{T}}$. In the follows, we talk about two standard linear methods for solving it.

I. Non-Homogeneous Solution In non-homogeneous method, one of nine elements in the Homographic matrix is assumed to be a fixed value. Usually, we let $h_9 = 1$ and stack the remaining eight elements into a vector $\boldsymbol{h} = (h_1, h_2, h_3, h_4, h_5, h_6, h_7, h_8)^{\mathrm{T}}$.

For the i-th pair of points, we have

$$\begin{aligned} (u_i, v_i, 1, 0, 0, 0, -u'_i u_i, -u'_i v_i)\boldsymbol{h} &= u'_i \\ (0, 0, 0, u_i, v_i, 1, -v'_i u_i, -v'_i v_i)\boldsymbol{h} &= v'_i \end{aligned} \tag{2.10}$$

From (2.10), given $n (n \geq 4)$ pairs of correspondences, we can have the following matrix equation:

$$\boldsymbol{A}\boldsymbol{h} = \boldsymbol{b} \tag{2.11}$$

where $\boldsymbol{A} = \begin{bmatrix} u_1 & v_1 & 1 & 0 & 0 & 0 & -u'_1 u_1 & -u'_1 v_1 \\ 0 & 0 & 0 & u_1 & v_1 & 1 & -v'_1 u_1 & -v'_1 v_1 \\ \vdots & \vdots & \vdots & \vdots & \vdots & \vdots & \vdots & \vdots \\ u_n & v_n & 1 & 0 & 0 & 0 & -u'_n u_n & -u'_n v_n \\ 0 & 0 & 0 & u_n & v_n & 1 & -v'_n u_n & -v'_n v_n \end{bmatrix}$

and $\boldsymbol{b} = (u'_1 \quad v'_1 \quad \cdots \quad u'_n \quad v'_n)^{\mathrm{T}}$.

Equation (2.11) is a standard linear equation system. There are many ways for solving it, such as Gaussian Elimination, LU decomposition and Jacobi iteration method. Once the vector $\boldsymbol{h}$ is obtained, the Homographic matrix can be got by unstacking this vector.

II. Homogeneous Solution In this case, we can stack the Homography matrix into a 9-vector as $\boldsymbol{h} = (h_1, h_2, h_3, h_4, h_5, h_6, h_7, h_8, h_9)^{\mathrm{T}}$.

For the i-th pair of points, we have

$$
\begin{aligned}
(u_i, v_i, 1, 0, 0, 0, -u'_i u_i, -u'_i v_i, u'_i)\boldsymbol{h} &= 0 \\
(0, 0, 0, u_i, v_i, 1, -v'_i u_i, -v'_i v_i, v'_i)\boldsymbol{h} &= 0
\end{aligned}
\tag{2.12}
$$

From (2.12), given $n(n \geq 4)$ pairs of correspondences, we can obtain the following matrix equation:

$$
\boldsymbol{A}\boldsymbol{h} = \boldsymbol{0} \tag{2.13}
$$

Where $\boldsymbol{A} = \begin{bmatrix} u_1 & v_1 & 1 & 0 & 0 & 0 & -u'_1 u_1 & -u'_1 v_1 & -u'_1 \\ 0 & 0 & 0 & u_1 & v_1 & 1 & -v'_1 x_1 & -v'_1 v_1 & -v'_1 \\ \vdots & \vdots & \vdots & \vdots & \vdots & \vdots & \vdots & \vdots & \vdots \\ u_n & u_n & 1 & 0 & 0 & 0 & -u'_n u_n & -u'_n v_n & -u'_n \\ 0 & 0 & 0 & u_n & v_n & 1 & -v'_n u_n & -v'_n v_n & -v'_n \end{bmatrix}$.

Let $\boldsymbol{Q} = \boldsymbol{A}^{\mathrm{T}}\boldsymbol{A}$. Using eigenvalue decomposition, the solution for the vector $\boldsymbol{h}$ can be determined by the eigenvector corresponding to the smallest eigenvalue of $\boldsymbol{Q}$.

Remark

1. The non-homogeneous solution has the disadvantage that poor estimation is obtained if the chosen element should actually have the value zero or close to zero. The homogeneous one overcomes this disadvantage.
2. Appropriate choice of the methods will provide convenience in different cases. For example, we will use the non-homogeneous solution when evaluating the computational complexity in Chap. 5, and the homogeneous solution will be used when doing error analysis in Chap. 4.

2.4.4 Fundamental Matrix

The homography matrix describes the mutual relationship among an arbitrary 3D planar patch, its images and the characteristic of the vision system, while the fundamental matrix encapsulates the intrinsic epipolar geometry. It is independent of the scene structure, and only depends on the camera's internal parameters and relative pose.

Mathematically, the fundamental matrix is a 3×3 matrix and the rank is 2. Since it is a singular matrix, there are many different parameterizations. For example, we can express one row (or column) of the fundamental matrix as the linear combination of the other two rows (or columns). As a result, there are many different approaches for estimating this matrix. We will next show a simple homogeneous solution.

Assuming an arbitrary point $\boldsymbol{M}$ in the scene, the corresponding image pixels are denoted by $\boldsymbol{m}_c$ and $\boldsymbol{m}_p$ in Fig. 2.13, then we have

$$
\boldsymbol{m}_c^{\mathrm{T}} \boldsymbol{F} \boldsymbol{m}_p = 0 \tag{2.14}
$$

where $\boldsymbol{F}$ is called the fundamental matrix.

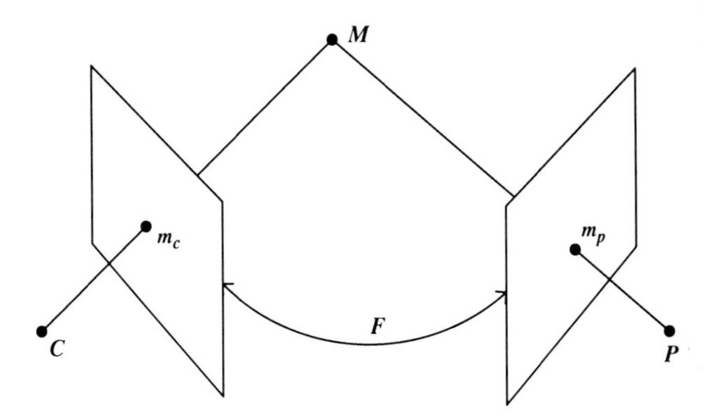

Fig. 2.13 An illustration for fundamental matrix

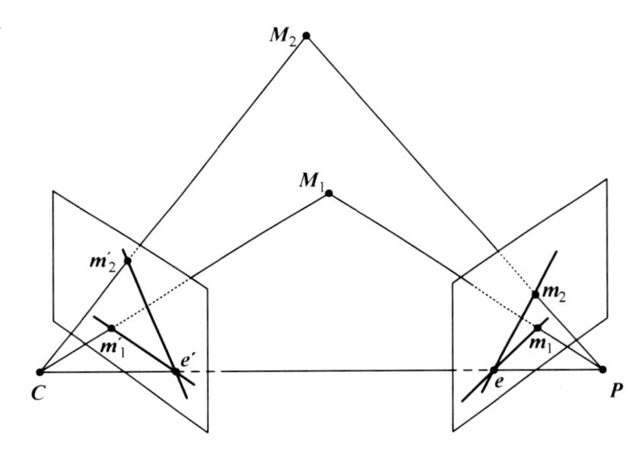

Fig. 2.14 Epipolar Geometry in a vision system

From (2.14), each point pair provides one constraint on the fundamental matrix. Let $\boldsymbol{F} = \begin{bmatrix} f_1 & f_2 & f_3 \\ f_4 & f_5 & f_6 \\ f_7 & f_8 & f_9 \end{bmatrix}$. Its row-first vector is $\boldsymbol{f} = (f_1, f_2, f_3, f_4, f_5, f_6, f_7, f_8, f_9)^{\mathrm{T}}$.

Let the coordinates of $\boldsymbol{m}_c$ and $\boldsymbol{m}_p$ be set as in the previous section. Rearranging (2.14), we have

$$[uu', uv', u, vu', vv', v, u', v', 1]\boldsymbol{f} = 0 \tag{2.15}$$

Similarly, given $n(n \geq 8)$ pairs of correspondences, we can obtain the following matrix equation:

$$\boldsymbol{Af} = \boldsymbol{0} \tag{2.16}$$

where $\boldsymbol{A}$ represents the coefficient matrix.

The solution for the vector $\boldsymbol{f}$ can be determined up to a scale factor using eigenvalue decomposition. And so is the fundamental matrix $\boldsymbol{F}$. The solution can be improved by nonlinear optimization with the following energy function:

$$E = \sum_i \left(d^2(\boldsymbol{m}_{ci}, \boldsymbol{F}\boldsymbol{m}_{pi}) + d^2(\boldsymbol{m}_{pi}, \boldsymbol{F}^{\mathrm{T}}\boldsymbol{m}_{ci})\right) \tag{2.17}$$

Remark The fundamental matrix describes the mutual relationship between any two images of the same scene. It has found various applications in computer vision. For example, given the projection of a scene point into one of the images the corresponding point in the other image is constrained to a line, helping the search of feature correspondences. This is generally termed as epipolar geometry (Fig. 2.14). With the fundamental matrix a projective reconstruction can be immediately obtained. We will discuss its use in a catadioptric camera system in Chap. 6.

Chapter 3
Static Calibration

The camera calibration is an important step towards structure from motion, automobile navigation and many other vision tasks. It mainly refers to the determination of camera intrinsic parameters, e.g. the focal lengths, the skew factor and the principal point. In practice, a good calibration algorithm should be robust, computational efficient and as convenient as possible for implementing. In this chapter, we first introduce the basic theory and methodology for determining camera intrinsic parameters. Then three different planar patterns are designed for camera calibration based on the theories. Finally, camera distortion correction is discussed. We will show that these approaches satisfy the above requirements. Besides those theoretic investigations, we have presented some examples and experimental results to demonstrate their applications.

3.1 Calibration Theory

In 3D projective geometry, the absolute conic (**AC**) is defined as a virtual point conic lying on the plane at infinity π_∞ (Fig. 3.1). It is a fixed conic under projective transformation. Algebraically, the absolute conic corresponds to a conic with matrix $\boldsymbol{\omega}_\infty = \mathbf{I}$, such that for an arbitrary point $\boldsymbol{M}_\infty$ on the conic, the identity $\boldsymbol{M}_\infty^{\mathrm{T}} \boldsymbol{\omega}_\infty \boldsymbol{M}_\infty = 0$ always holds.

Let $\widetilde{\boldsymbol{M}}_\infty$ be the homogeneous representation of $\boldsymbol{M}_\infty$, which means $\widetilde{\boldsymbol{M}}_\infty = [x\ y\ z\ 0]^{\mathrm{T}}$ if we set $\boldsymbol{M}_\infty = [x\ y\ z]^{\mathrm{T}}$. Let $\widetilde{\boldsymbol{m}}_\infty$ be its corresponding imaged point. Let the intrinsic and extrinsic parameters of the camera be respectively $\boldsymbol{K}_c$ and $\boldsymbol{R}_c$, $\boldsymbol{t}_c$. From (2.3), we have

$$\widetilde{\boldsymbol{m}}_\infty = \beta \boldsymbol{K}_c [\boldsymbol{R}_c \quad \boldsymbol{t}_c] \widetilde{\boldsymbol{M}}_\infty \tag{3.1}$$

Then we obtain

$$\widetilde{\boldsymbol{m}}_\infty^{\mathrm{T}} \boldsymbol{K}_c^{-\mathrm{T}} \boldsymbol{K}_c^{-1} \widetilde{\boldsymbol{m}}_\infty = \beta^2 \widetilde{\boldsymbol{M}}_\infty^{\mathrm{T}} [\boldsymbol{R}_c \quad \boldsymbol{t}_c]^{\mathrm{T}} [\boldsymbol{R}_c \quad \boldsymbol{t}_c] \widetilde{\boldsymbol{M}}_\infty = \beta^2 \boldsymbol{M}_\infty^{\mathrm{T}} \boldsymbol{\omega}_\infty \boldsymbol{M}_\infty = 0 \tag{3.2}$$

The above equation describes the relationship between $\widetilde{\boldsymbol{M}}_\infty$ and its image point $\widetilde{\boldsymbol{m}}_\infty$. In other words, it indicates that the image of the absolute conic (**IAC**) is a conic

B. Zhang, Y. F. Li, *Automatic Calibration and Reconstruction for Active Vision Systems*,
Intelligent Systems, Control and Automation: Science and Engineering,
DOI 10.1007/978-94-007-2654-3_3,

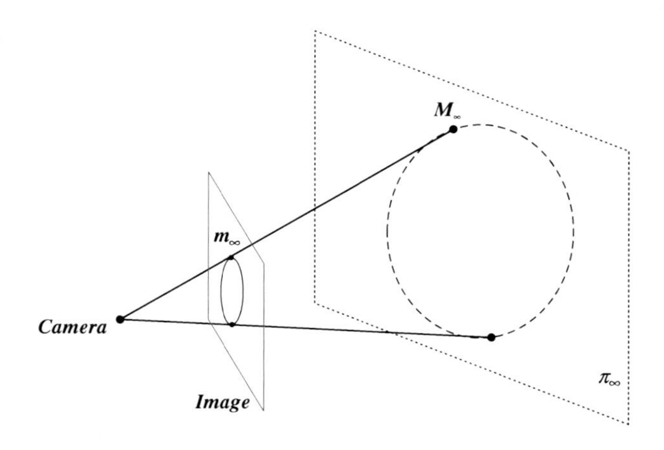

Fig. 3.1 The absolute conic and its image

with matrix $\boldsymbol{\Theta} = \boldsymbol{K}_c^{-T}\boldsymbol{K}_c^{-1}$, which is directly related to the intrinsic parameters of the camera but independent of the extrinsic parameters. If this matrix is found, we can extract $\boldsymbol{K}_c^{-1}$ by Cholesky factorization and hence find the internal camera matrix $\boldsymbol{K}_c$ by matrix inverse. This is the basic calibration theory for the internal parameters of the camera using the image of the absolute conic. From (3.2), we can derive the constraints on the image of absolute conic given the point $\tilde{\boldsymbol{m}}_\infty$. Consequently, the camera calibration problem becomes how to find the point $\boldsymbol{M}_\infty$ or $\tilde{\boldsymbol{m}}_\infty$ and hence the matrix $\boldsymbol{\Theta}$.

On the other hand, the intersection of the line at infinity and a circle is named as the circular points. With a few mathematical manipulations, the circular points can be expressed as $\boldsymbol{J} = (1, j, 0,0)^{\mathrm{T}}$ and $\overline{\boldsymbol{J}} = (1, -j, 0,0)^{\mathrm{T}}$, here $j = \sqrt{-1}$. Obviously, they are a pair of conjugate points and have nothing to do with the center and radius of the circle. Furthermore, it is easy to verify that the circular points belong to the absolute conic, which means $\boldsymbol{J}^{\mathrm{T}}\omega_\infty\boldsymbol{J} = 0$ and $\overline{\boldsymbol{J}}^{\mathrm{T}}\omega_\infty\overline{\boldsymbol{J}} = 0$ always hold.

Now, assuming the imaged points of the circular points are $\boldsymbol{m}_J$ and $\boldsymbol{m}_{\bar{J}}$, from (3.2) we have

$$\boldsymbol{m}_J^{\mathrm{T}}\boldsymbol{\Theta}\boldsymbol{m}_J = 0 \quad \text{and} \quad \boldsymbol{m}_{\overline{J}}^{\mathrm{T}}\boldsymbol{\Theta}\boldsymbol{m}_{\overline{J}} = 0 \tag{3.3}$$

In projective transformation, $\boldsymbol{m}_J$ and $\boldsymbol{m}_{\bar{J}}$ are also a pair of conjugate points. So (3.3) only provides the following two constraints on the intrinsic parameters

$$\boldsymbol{Re}\left(\boldsymbol{m}_J^{\mathrm{T}}\boldsymbol{\Theta}\boldsymbol{m}_J\right) = 0 \quad \text{and} \quad \boldsymbol{Im}\left(\boldsymbol{m}_J^{\mathrm{T}}\boldsymbol{\Theta}\boldsymbol{m}_J\right) = 0 \tag{3.4}$$

where: ***Re*** and ***Im*** denote the real and imaginary part of the expressions respectively.

If the camera model has two variable intrinsic parameters, two such equations are required to solve for them. Hence, a single image is sufficient in this case. Similarly, we need two or more images to provide enough constraints if more than two variables are considered. In what follows, we suggest three new different planar patterns for estimating the imaged circular points.

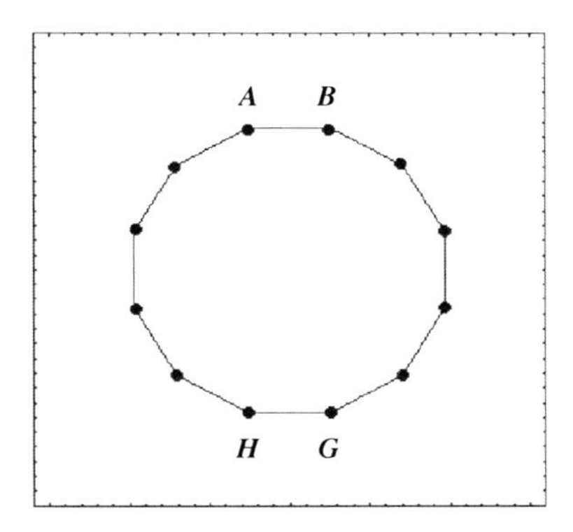

Fig. 3.2 The proposed planar pattern from [185]

3.2 Polygon-based Calibration

3.2.1 Design of the Planar Pattern

From (3.4), the key step for camera calibration is to find the projections of the circular points. How to design a planar pattern and achieve the required projections from the pattern conveniently is of critical importance here. For this purpose, we design one kind of planar pattern containing equilateral polygons with $2n$ sides ($n > 2$), of which the vertexes can be used to approximate the circle while the edges used to compute the line at infinity. In the image, they respectively correspond to the imaged ellipse and the vanishing line. Consequently, the intersection points of the ellipse and the vanishing line provide the projections of the circular points.

We can see that the designed planar pattern satisfies the requirements of camera calibration. In practice, the larger the number n, the more accurately this polygon can generally approximate the circle and the line at infinity in the presence of noise. However, a large n increases the computational load in the approximation and the complexity in the image processing. From our experiments, we find that an acceptable compromise is achieved when $n = 6$. This results in a planar pattern of an equilateral dodecagon, as shown in Fig. 3.2.

3.2.2 Solving the Vanishing Line

The equilateral dodecagon has twelve sides grouped into six pairs of parallel lines. They intersect at six ideal points that lie on a common line, namely the line at infinity. According to the principle of projection, their projections will intersect at six vanishing points that lie on the vanishing line in the image (Fig. 3.3). For example, if $\boldsymbol{l}_{ab}$ and $\boldsymbol{l}_{hg}$ are the projections of $\overline{\boldsymbol{AB}}$ and $\overline{\boldsymbol{HG}}$, we can obtain a vanishing point by $\boldsymbol{m}_v = \boldsymbol{l}_{ab} \times \boldsymbol{l}_{hg}$.

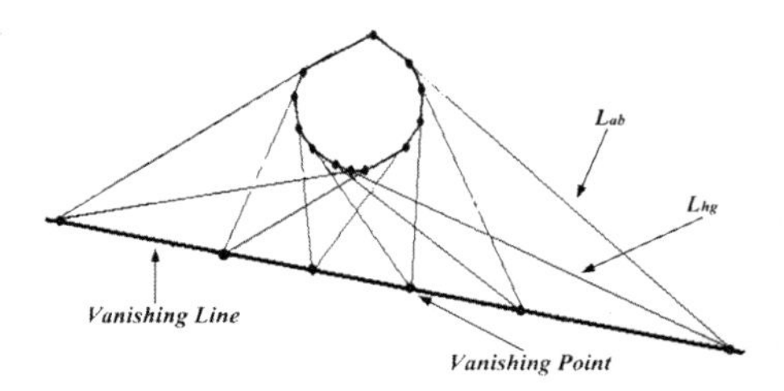

Fig. 3.3 Vanishing points and vanishing line

Consequently, we can determine the vanishing line by choosing any two among the six vanishing points or using all in the least squares sense. Let the obtained line vector be denoted by $\boldsymbol{l}_v$.

3.2.3 Solving the Projection of a Circle

The equilateral dodecagon has a circum-circle passing through its twelve vertexes. In general, the projection of the circum-circle in the image is an ellipse. According to the principle of projection, the projections of these vertexes lie on the ellipse. Hence, they can be used to approximate the ellipse, i.e. the projection of the circle.

It is well known that an ellipse can be represented by a 3×3 symmetric matrix. Without loss of generality, let $\boldsymbol{E} = \begin{bmatrix} a & d & e \\ d & b & f \\ e & f & c \end{bmatrix}$ denote the ellipse and $\boldsymbol{m}$ be a point on it. We have $\boldsymbol{m}^{\mathrm{T}}\boldsymbol{E}\boldsymbol{m} = 0$ which provides one constraint on the ellipse. Hence five points are sufficient to determine it up to a scale factor. To improve the reliability, we can stack all the constraints into the following linear systems

$$\boldsymbol{A}\boldsymbol{x} = 0 \tag{3.5}$$

where

$$\boldsymbol{A} = \begin{bmatrix} u_1^2 & v_1^2 & 1 & 2u_1v_1 & 2u_1 & 2v_1 \\ u_2^2 & v_2^2 & 1 & 2u_2v_2 & 2u_2 & 2v_2 \\ \vdots & \vdots & \vdots & \vdots & \vdots & \vdots \\ u_{12}^2 & v_{12}^2 & 1 & 2u_{12}v_{12} & 2u_{12} & 2v_{12} \end{bmatrix}, \quad \boldsymbol{x} = (a \quad b \quad c \quad d \quad e \quad f)^{\mathrm{T}}$$

and $\quad \boldsymbol{m}_i = (u_i, v_i)^{\mathrm{T}}$.

The above equation can be solved by SVD or eigenvalue decomposition to obtain the ellipse vector $\boldsymbol{x}$, and so is the ellipse matrix $\boldsymbol{E}_c$.

3.2.4 Solving the Projection of Circular Point

Once the vanishing line $\boldsymbol{l}_v$ and the ellipse $\boldsymbol{E}_c$ have been estimated, the projection of circular point can be solved by their intersections.

Mathematically, it is represented as $\begin{cases} \boldsymbol{l}_v^{\mathrm{T}} \boldsymbol{m}_J = 0 \\ \boldsymbol{m}_J^{\mathrm{T}} \boldsymbol{E}_c \boldsymbol{m}_J = 0 \end{cases}$.

This consists of one linear equation and one quadratic equation and gives two solutions, representing the two intersections. In general, they should be conjugate with each other.

3.2.5 Algorithm

Summarily, the implementation procedure for camera calibration can be divided into the following six steps:

1. Printing an equilateral dodecagon on a sheet paper to make the planar pattern;
2. Moving either the planar pattern or the camera to take a few (at least three) images under different locations;
3. For each image, extracting the vertexes of the equilateral dodecagon and then approximating the ellipse and the vanishing line using these vertexes;
4. Calculating the projection of circular point by intersecting the ellipses and the vanishing lines;
5. Constructing the linear constraints on $\boldsymbol{\Theta}$ according to (3.4) and solving it by singular value decomposition;
6. Extracting $\boldsymbol{K}_c^{-1}$ by Cholesky factorization of $\boldsymbol{\Theta}$, then inversing it to obtain the camera matrix $\boldsymbol{K}_c$.

The calibration results can be improved using nonlinear optimization with the information from all the images.

3.2.6 Discussion

In practice, a good camera calibration algorithm should have two important features. One is that the calibration pattern should be easy to make and convenient to manipulate. The other is that the calibration process should be as simple as possible. In this algorithm, the position and the metric size of the equilateral dodecagon can be arbitrary. And the point matching between the image and the pattern is no longer needed when working. It is sufficient to find those vertexes in each image. These characteristics greatly simplify the pattern making and the image processing. So it satisfies the two standards.

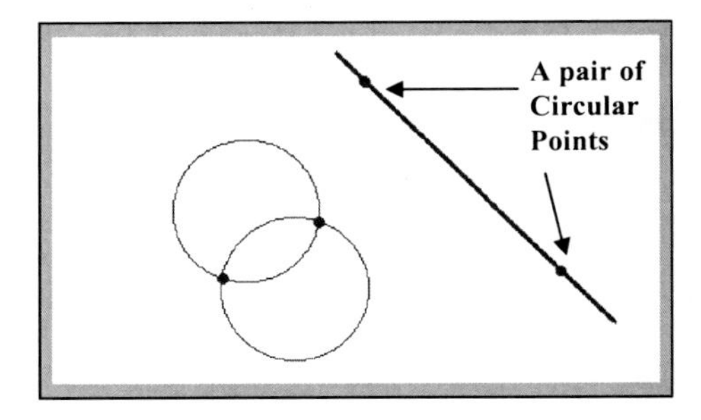

Fig. 3.4 Coplanar circles and the circular points from [186]

3.3 Intersectant-Circle-based Calibration

3.3.1 Planar Pattern Design

In the previous section, the key problem is how to efficiently find the image of the circular points, which represents the intersection of the vanishing line and the ellipse. Here, we have two important observations: one is that the circular point has nothing to do with the position and metric size of the circle in the planar pattern; and the other is that the intersection of two circles is an invariant property under projective transformation. Therefore, we arrive at the conclusion that all coplanar circles have the same intersection points with the line at infinity. This means that they all pass through the circular points or the circular points can be estimated by the intersecting operation of those circles (Fig. 3.4).

It is known that each circle can be algebraically denoted by a quadric equation with two variables. Given two circles, we have two such equations. In the complex space, they should have four roots corresponding to the four intersection points. If we take a look in details, we can find that two of the four are real solutions and the rest two are a pair of conjugate solutions. The latter should be the circular points based on the above analysis. Hence, we can design a planar pattern containing two circles for the static calibration.

3.3.2 Solution for The Circular Point

In the image space, each circle is imaged as an ellipse and its quadratic equation with two variables can be approximated by the pixels. Figure 3.5 gives three images of the pattern in different positions. The process for fitting those ellipses is exactly the same as what we have done in Sect. 3.2.3. Now, the problem becomes how to solve the quadratic equation system to obtain the image of circular points.

We will show a programmable strategy to solve it here. It consists of three steps: firstly take one variable, say v in (3.6), as a constant and solve the other variable $u = f(v)$; secondly substitute the roots of u into (3.7) to obtain a standard quartic equation and solve it for v; lastly back substitute the roots of v to get the solutions of u. The details are given as what follows.

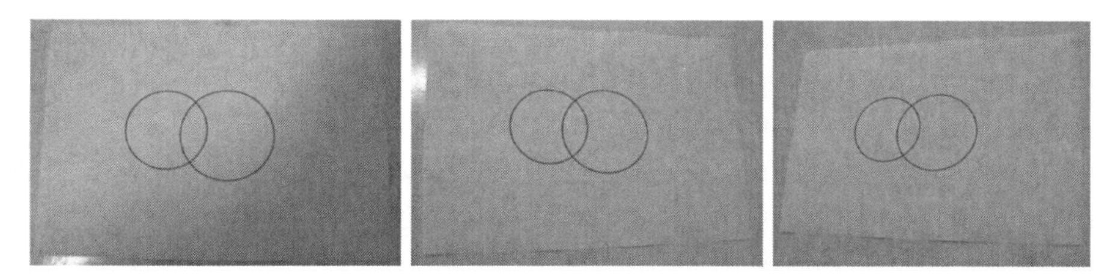

Fig. 3.5 Three images of the pattern

Supposing the equations for the two ellipses be

$$a_1u^2 + b_1v^2 + c_1uv + d_1u + e_1v + f_1 = 0 \tag{3.6}$$

and

$$a_2u^2 + b_2v^2 + c_2uv + d_2u + e_2v + f_2 = 0 \tag{3.7}$$

Equations (3.6) and (3.7) provide two constraints on the intersection points, i.e. their elements u and v. From (3.6), u can be represented by v:

$$u = \frac{-c_1v - d_1}{(2a_1)} \pm \frac{\left(\left(c_1^2 - 4a_1b_1\right)v^2 + (2c_1d_1 - 4a_1e_1)\,v + d_1^2 - 4a_1f_1\right)^{\frac{1}{2}}}{(2a_1)} \tag{3.8}$$

Substituting (3.8) into (3.7), we can derive a quartic equation with one variable v as the following form:

$$av^4 + bv^3 + cv^2 + dv + e = 0 \tag{3.9}$$

Let $y = v - \frac{b}{4a}$. Rearranging Eq. (3.9), we obtain

$$y^4 + py^2 + qy + r = 0 \tag{3.10}$$

There are many methods for solving (3.10), e.g. method of completing square, factorization method and method of parameter increment. With the solution of variable y and hence u, the other element v can be determined by back substitutions. Consequently, the conjugate complex roots become the solutions for the projection of the circular points, from which two constraints with the form of (3.4) can be established for each image. With three or more images, we have six or more such constraints. Stacking them, then the camera calibration can be carried out following the algorithm given in Sect. 3.2.5.

Remark In this section, the two circles in the pattern intersect with each other. This simplifies the calibration process. Another advantage of this technique is that the identification of the vanishing line, which can be computationally expensive and sometimes even impossible, is no longer required. When solving the two quadratic equations, the two real solutions correspond to the real intersection points while the conjugate complex solutions are what we need.

In fact, the positions of the circles can be arbitrary, e.g. being concentric circles or separating from each other completely. In these cases, a verification procedure is required to find the true projection of circular points since all the solutions are conjugate complex points. This may be laborious and tedious in some applications. Furthermore, this technique involves the solution a quatic equation. It is well known that the root finding for high order polynomials is an ill-conditioned problem, where slightest perturbations in the coefficients of the equation can cause the solutions to not only change drastically but in some cases become complex.

In the next section, we will treat this problem from a different viewpoint and suggest a convenient solution in case of concentric circles.

3.4 Concentric-Circle-based Calibration

In this calibration algorithm, the key problem is how to find the projection of the center for the concentric circles. In the literatures, there are some solutions for this problem. With the property of projected coplanar circles, Zhao et al [176] firstly computed the lines connecting the images of circle centers, and then those centers can be determined from the intersection of the computed lines. This technique requires at least 2×2 array circles and hence the pattern is hard to find in our surroundings. To decrease the number of required circles, Jiang and Quan [177] proposed a constructive algorithm based on a simple geometric observation as illustrated in Fig. 3.6: Given a circle in this figure, they tried to reach the center $\boldsymbol{O}$ of the circle from an arbitrary point $\boldsymbol{P}$ inside the circle. Firstly drawing an arbitrary line through the point $\boldsymbol{P}$, and taking the midpoint $\boldsymbol{P1}$ of the chord generated by the line. Then, repeating the same by drawing another arbitrary line through the point $\boldsymbol{P1}$, and taking the midpoint $\boldsymbol{P2}$ of the chord generated by the new line, and so on. The observation is that

$$|\boldsymbol{OP}_1| > |\boldsymbol{OP}_2| > \cdots > |\boldsymbol{OP}_{i-1}| > |\boldsymbol{OP}_i|$$

Therefore, continuing this construction should result in a sequence points that will strictly converge at the center of the circle as long as each time a different line is drawn. This method is geometrically instinctive and interesting. However, there are two difficulties when implementation. One is how to provide an appropriate convergence criterion that will end the construction process. The other is how to design a computationally efficient algorithm to carry out that task. Furthermore, we find that the selection of a good start point and searching direction may have an important effect on the efficiency in some case.

In this section, we will talk about a simple method for finding the projection center of the concentric circles, in which the solution for this problem is formulated into a first-order polynomial eigenvalue problem by considering the pole-polar relationship in the image. The major advantage is that it is an analytic algorithm and no iterations are needed.

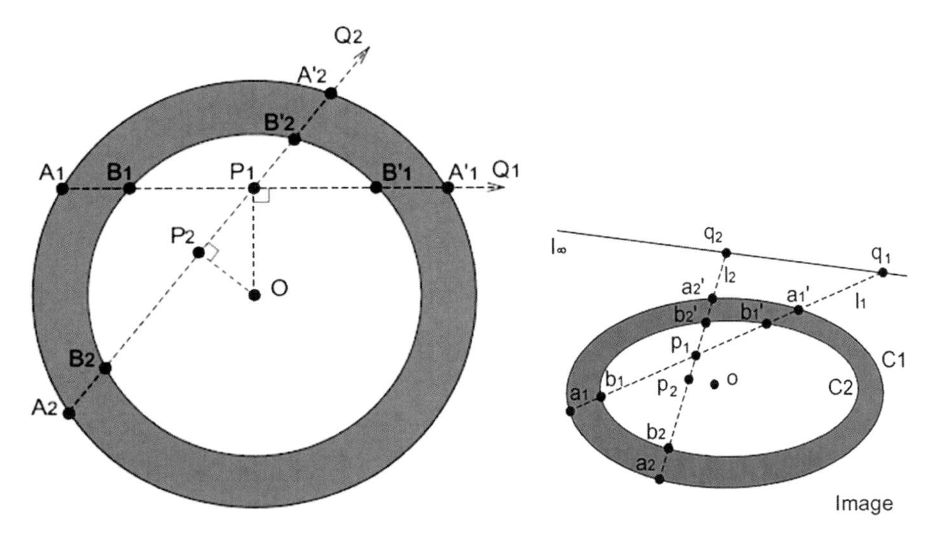

Fig. 3.6 Construction of a sequence of points ***P1***, ***P2***, ... strictly convergent to the center of the circle from any point ***P***

3.4.1 Some Preliminaries

In Euclidean geometry, the angle between two lines is computed from the dot product of their normal. However, this is not the case in the image space under projective transformation. The following result provides the solution, which is invariant to the projective transformation.

Result 3.1 Let $\boldsymbol{v}_1$ and $\boldsymbol{v}_2$ denote the vanishing points of two lines in an image, and let $\boldsymbol{\Theta}$ be the image of the absolute conic. If θ is the angle subtended by the two lines, then we have

$$\cos\theta = \frac{\boldsymbol{v}_1^{\mathrm{T}}\boldsymbol{\Theta}\boldsymbol{v}_2}{\sqrt{\boldsymbol{v}_1^{\mathrm{T}}\boldsymbol{\Theta}\boldsymbol{v}_1}\sqrt{\boldsymbol{v}_2^{\mathrm{T}}\boldsymbol{\Theta}\boldsymbol{v}_2}} \tag{3.11}$$

In projective geometry, a point $\boldsymbol{x}$ and a conic $\boldsymbol{C}$ define a line $\boldsymbol{l} = \boldsymbol{C}\boldsymbol{x}$ (Fig. 3.7). The line $\boldsymbol{l}$ is called the polar of $\boldsymbol{x}$ with respect to $\boldsymbol{C}$, and the point $\boldsymbol{x}$ is the pole of $\boldsymbol{l}$. This is the pole-polar relationship.

Result 3.2 Let $\boldsymbol{C}$ be a circle. With the pole-polar relationship, $\boldsymbol{x}$ should be the circle center if and only if $\boldsymbol{l}$ is the line at infinity of the supporting plane. In the image space, $\boldsymbol{C}$ is imaged into an ellipse, $\boldsymbol{l}$ into a vanishing line and $\boldsymbol{x}$ into the image of circle center.

This result can be briefly showed as what follows. Let $\boldsymbol{x} = [x_0, y_0, 1]^{\mathrm{T}}$ be the circle center and r the radius. Then the circle $\boldsymbol{C}$ is represented in matrix form

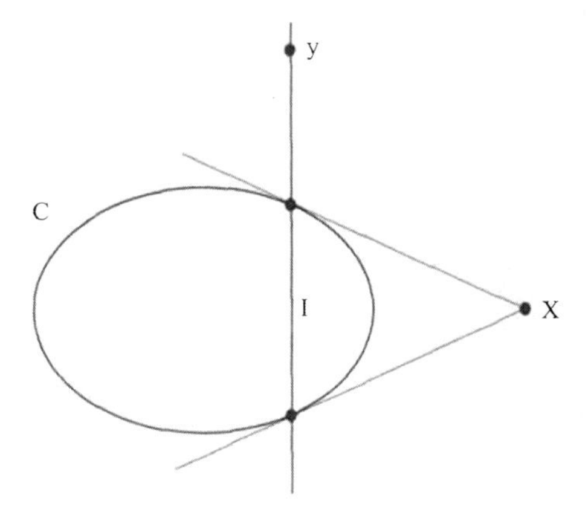

Fig. 3.7 The pole-polar relationship from Hartley and Zisserman [150]

as $\begin{bmatrix} 1 & 0 & -x_0 \\ 0 & 1 & -y_0 \\ -x_0 & -y_0 & x_0^2+y_0^2-r_0^2 \end{bmatrix}$. By mathematical manipulation, it is easy to obtain $\boldsymbol{C}\boldsymbol{x} \sim [0 \quad 0 \quad 1]^{\mathrm{T}}$, i.e. the line at infinity $\boldsymbol{l}$. On the other hand, we have $\boldsymbol{C}^{-1}\boldsymbol{l} \sim [x_0 \quad y_0 \quad 1]^{\mathrm{T}}$, i.e. the circle center. Here, the notation $\sim$ means equality up to a nonzero scale factor. So we have the conclusion that the polar of the circle will be the line at infinity if its center is the pole, and vice versa.

Proposition 3.1 *Let $\boldsymbol{C}_1$ and $\boldsymbol{C}_2$ be two concentric circles and their projections in the image be $\boldsymbol{E}_1$ and $\boldsymbol{E}_2$ respectively. Then the matrix $s_2\boldsymbol{E}_2 - s_1\boldsymbol{E}_1$ has a pair of identical eigenvalues, which are different from the third one.*

Proof Without loss of generality, we assume the center of the circles be $[x_0, y_0]^{\mathrm{T}}$ and the radiuses be r_1, r_2. Then we have

$$\boldsymbol{C}_1 = \begin{bmatrix} 1 & 0 & -x_0 \\ 0 & 1 & -y_0 \\ -x_0 & -y_0 & x_0^2+y_0^2-r_1^2 \end{bmatrix} \quad \text{and} \quad \boldsymbol{C}_2 = \begin{bmatrix} 1 & 0 & -x_0 \\ 0 & 1 & -y_0 \\ -x_0 & -y_0 & x_0^2+y_0^2-r_2^2 \end{bmatrix}.$$

Let the projection matrix be $\boldsymbol{H}$. According to the transformation of conics, we obtain $s_1\boldsymbol{E}_1 = \boldsymbol{H}^{-\mathrm{T}}\boldsymbol{C}_1\boldsymbol{H}^{-1}$ and $s_2\boldsymbol{E}_2 = \boldsymbol{H}^{-\mathrm{T}}\boldsymbol{C}_2\boldsymbol{H}^{-1}$, where s_1 and s_2 are nonzero scale factors.

$$\begin{aligned} s_2\boldsymbol{E}_2 - s_1\boldsymbol{E}_1 &= \boldsymbol{H}^{-\mathrm{T}}\boldsymbol{C}_2\boldsymbol{H}^{-1} - \boldsymbol{H}^{-\mathrm{T}}\boldsymbol{C}_1\boldsymbol{H}^{-1} \\ &= \boldsymbol{H}^{-\mathrm{T}}(\boldsymbol{C}_2 - \boldsymbol{C}_1)\boldsymbol{H}^{-1} \\ &= \boldsymbol{H}^{-\mathrm{T}} \begin{bmatrix} 0 & 0 & 0 \\ 0 & 0 & 0 \\ 0 & 0 & r_1^2 - r_2^2 \end{bmatrix} \boldsymbol{H}^{-1} \end{aligned}$$

In the above matrix equation, the element $r_1^2 - r_2^2$ cannot be zero since the two circles have different radiuses. According to the property of similarity transformation, we

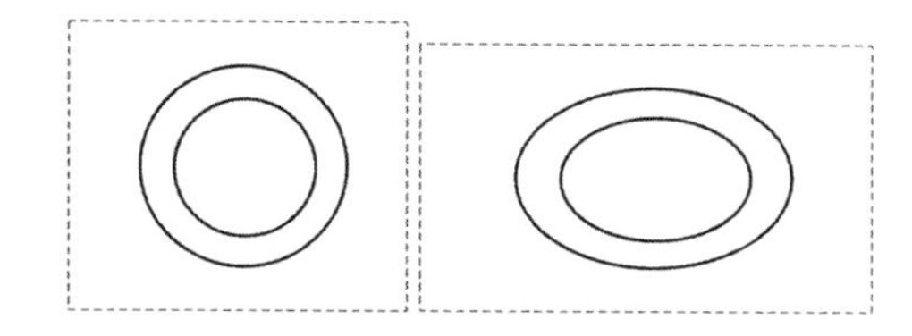

Fig. 3.8 The planar pattern and its image

have the conclusion that matrix $s_2 \boldsymbol{E}_2 - s_1 \boldsymbol{E}_1$ has a pair of identical eigenvalues, which are different from the third one. □

3.4.2 The Polynomial Eigenvalue Problem

The left figure in Fig. 3.8 presents the planar pattern containing two concentric circles, and the right illustrates its image. We assume that the two ellipses in the image have been estimated as what we have done in Sect. 3.2. Let them be denoted by $\boldsymbol{E}_1$ and $\boldsymbol{E}_2$ respectively. Let the image of the circle center be $\boldsymbol{c}$ and the vanishing line of its supporting plane be $\boldsymbol{l}$.

From Result 3.2, we have

$$\boldsymbol{l} = \lambda_1 \boldsymbol{E}_1 \boldsymbol{c} \tag{3.12}$$

and

$$\boldsymbol{l} = \lambda_2 \boldsymbol{E}_2 \boldsymbol{c} \tag{3.13}$$

where λ_1 and λ_2 are nonzero scale factors.

From (3.12) and (3.13), we obtain

$$\boldsymbol{E}_1 \boldsymbol{c} = \sigma \boldsymbol{E}_2 \boldsymbol{c} \tag{3.14}$$

where $\sigma = \lambda_2/\lambda_1$.

Rearranging (3.14),

$$(\sigma \boldsymbol{E}_2 - \boldsymbol{E}_1)\boldsymbol{c} = 0 \tag{3.15}$$

In (3.15), the unknowns include the center projection $\boldsymbol{c}$ and the scale factor σ. $\boldsymbol{E}_1$ and $\boldsymbol{E}_2$ are 3×3 symmetric matrix representing the two ellipses. This is known as a first-order polynomial eigenvalue problem (**PEP**).

There are many researches on the solutions of **PEP** in the literature, e.g. [179]. In practice, some efficient algorithms are available for solving it, e.g. MATLAB solves the **PEP** by the function named *polyeig()*. There are three eigenvalues obtained since the matrix size is 3×3. According to Proposition 3.1, two of them are identical and different from the third one. The correct solution for the center projection is the eigenvector corresponding to the third eigenvalue.

Once $\boldsymbol{c}$ is estimated, the vanishing line $\boldsymbol{l}$ can be recovered by using either (3.12) or (3.13).

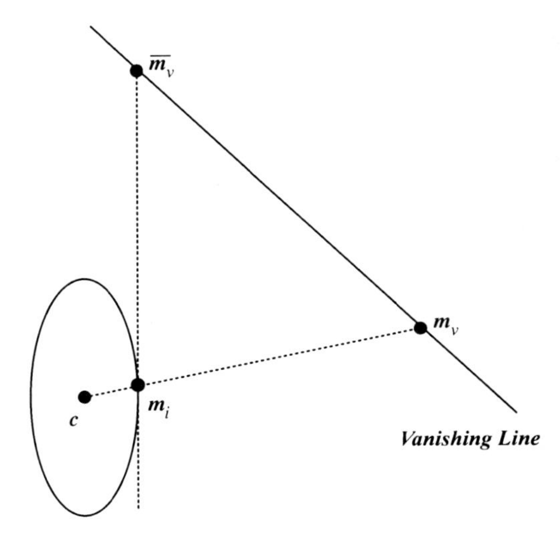

Fig. 3.9 Orthogonal relationship

3.4.3 *Orthogonality-Based Algorithm*

Once the vanishing line has been recovered, the intersection of this line with one of the two ellipses gives the image of the circular point. Hence, similar procedure as in Sect. 3.2.5 can be adopted for camera calibration. Alternatively, we present an orthogonality-based algorithm here.

It is well known that the line connect the circle center and a circle point is orthogonal to the tangential line through that point. This fact is depicted in Fig. 3.9. For any point $\boldsymbol{M}_i$ lie on the circle, the vanishing point $\boldsymbol{m}_v$ of the line connecting the point and the circle center can be computed by

$$\boldsymbol{m}_v = (\boldsymbol{c} \times \boldsymbol{m}_i) \times \boldsymbol{l} \tag{3.16}$$

where $\boldsymbol{m}_i$ is the image of $\boldsymbol{M}_i$.

Then the vanishing point $\overline{\boldsymbol{m}}_v$ of the tangential line passing through $\boldsymbol{M}_i$ is given by

$$\overline{\boldsymbol{m}}_v = (\boldsymbol{E} \times \boldsymbol{m}_i) \times \boldsymbol{l} \tag{3.17}$$

It should be noted that (3.16) and (3.17) are applicable for any of the two circles.

Obviously, $\boldsymbol{m}_v$ and $\overline{\boldsymbol{m}}_v$ are orthogonal to each other. According to Result 3.1, we have

$$\boldsymbol{m}_v^{\mathrm{T}} \boldsymbol{\Theta} \overline{\boldsymbol{m}}_v = 0 \tag{3.18}$$

Equation (3.18) provides one constraint on the image of the absolute conic. More constraints can be constructed with other points on the concentric circles. However, only two of them are independent given a single image, which means two variables can be solved. If more than two camera parameters are considered, we should move

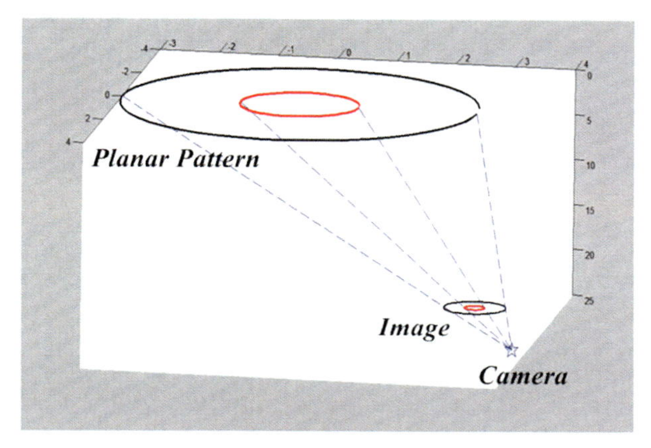

Fig. 3.10 One scenario of the simulated system

either the camera or the pattern to obtain more constraints. Once the image of the absolute conic $\boldsymbol{\Theta}$ is available, the camera intrinsic parameters can be recovered by Cholesky decomposition.

Summarily, here is the proposed calibration algorithm:

Step 1: Arbitrarily moving either the camera or the planar pattern to acquire three or more images;
Step 2: For each image, calculating the ellipse matrices using (3.5);
Step 3: Solving (3.15) to obtain the projection of the circle center, then recovering the vanishing line;
Step 4: Estimating the two orthogonal points $\boldsymbol{m}_v$ and $\overline{\boldsymbol{m}}_v$ from (3.16) and (3.17);
Step 5: Constructing the constraints on the absolute conic from (3.18) and stacking them all into a linear system.
Step 6: Solving the linear system by SVD to obtain $\boldsymbol{\Theta}$ and then recovering the camera matrix $\boldsymbol{K}$.

3.4.4 Experiments

3.4.4.1 Numerical Simulations

In the numerical simulations, we will test the accuracy and robustness of the suggested technique against different levels of Gaussian noise. Here, the camera matrix is assumed to be $\boldsymbol{K} = [1000, 2, 200; 0, 900, 100; 0, 0, 1]$. The radii of the concentric circles respectively be $r_1 = 1$ and $r_2 = 3$ while their center coincides with the origin of the world coordinate system for simplicity. The rotation matrix and translation vector are randomly chosen for each generated image. Figure 3.10 illustrates one of the simulated scenarios, which encodes the spatial relationship between the pattern, the image and the camera. We use TEN images to provide over-determined equations on the camera parameters. In each image, 20 pixels were used to approximate per

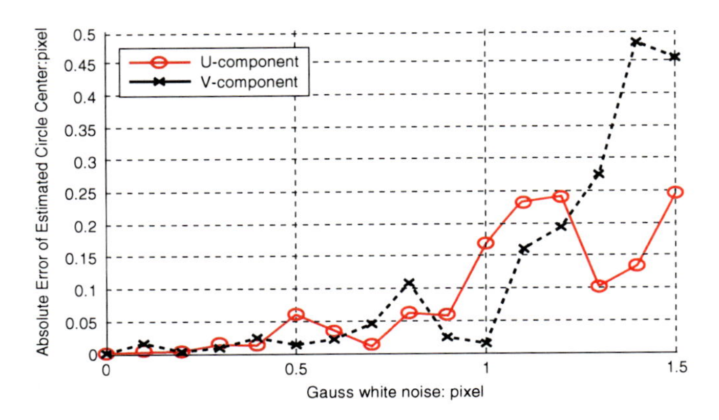

Fig. 3.11 Errors in estimated circle center vs. Gauss noise

ellipse. Gaussian noise with zero mean and standard deviation ranging from 0 to 1.5 pixels were added to those pixels.

In the first simulation, we investigated the accuracy of the estimation for the image of circle center since it is the key problem in our algorithm. The absolute error, which is defined as the difference between the true value and the estimated value, was recorded for U-component and V-component of the imaged circle center. In each level of noise, one randomly generated image was used to compute the circle center with the proposed technique. The results for all levels were shown in Fig. 3.11. We can see that the errors are exactly zeroes as expected when no noise is added, and become large along with the increased noise but within a very reasonable range, say less than 0.6 pixel in most cases.

In another simulation, we studied the calibration robustness of our algorithm against different levels of the noise. Here, the absolute estimation errors for the five elements in the camera matrix were recorded respectively. For each level of noise, we performed fifty random runs and averaged the obtained errors. Figure 3.12 presented the results for all levels of noise, in which the absolute errors of f_u and f_v are denoted by RED and GREEN while s, u_0 and v_0 by BLUE, YELLOW and BLACK. We observed that a satisfactory performance was obtained in this simulation. It is noticed that the errors for the last three elements are always smaller compared with the first two. When the noise becomes larger, the estimation of ellipse center is unstable which will lead to the wide variation in f_u and f_v. In fact, the error percent for all of them is less than 2%. Therefore, the robustness of our algorithm is acceptable.

3.4.4.2 Real Image Experiment

In the real image experiment, the vision system consists of just one USB camera (MVC3000 F from Microview Science & Technology Co.) with resolution 1280 × 1024. A computer monitor with two concentric circles acts as the planar

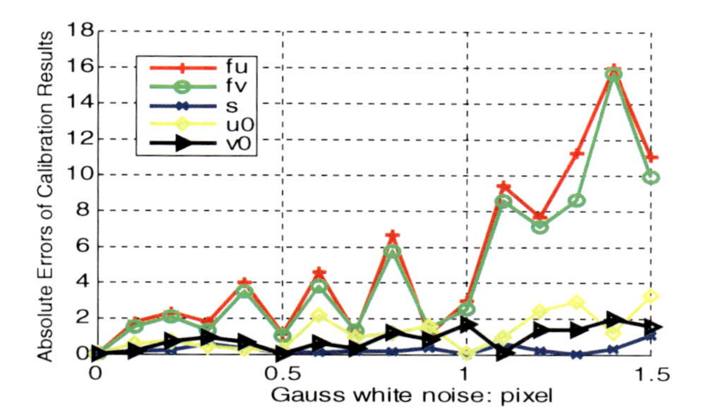

Fig. 3.12 The absolute errors for camera calibration results

pattern in a fixed position. The system is held by hand and moved arbitrarily to simulate free motions.

Figure 3.13 shows ten images taken by the camera when moving in our lab arbitrarily. Twenty pixels are randomly selected per ellipse in each image. Then the image of circle center, i.e. the ellipse center is estimated with the suggested technique. Here, the detected ellipses and those ellipse centers are illustrated by YELLOW edges and RED crosses. For comparison, the algorithm proposed in [177] is also implemented. We find that the estimated results are very close to each other in both methods (It should be noted that location optimization is involved in [177].). But the averaging time taken for processing each image is about 0.05 s in the former while around 0.12 s is needed in the latter. Hence, it is obvious that our technique is computationally more efficient.

With the detected ellipse center and vanishing line for each image, we can then perform the calibration algorithms. The calibrated results for the camera matrix are: $f_u = 3843.8$, $f_v = 4111$, $s = 15.7$, $u_0 = 1119.1$, $v_0 = 963$. Once the camera's intrinsic parameters are known, their extrinsic parameters can be computed with the knowledge of perpendicular edges of the computer monitor. To visually show the calibration results, we assume that the first camera position is placed at the origin of the world frame with coordinates [0, 0, 0]. Then the position is transferred to the second using the extrinsic parameters. This process is repeated until the last one. Finally, we obtain the whole motion trajectory as shown in Fig. 3.14. From this figure, we can see that the calibration results agree with the track of real motions.

Our last evaluation of this experiment is to carry out 3D reconstruction of the imaged scene using the fifth and sixth images. Here, Harris corner detection algorithm is firstly used to extract the features on the fifth image. Then those features are matched in the sixth image. Figure 3.15a and 3.15b gives the corresponding points in the two images, in which 28 feature points are plotted by RED ' + ' together with their sequence numbers. Figure 3.15c shows a portion of the fifth image for clarity.

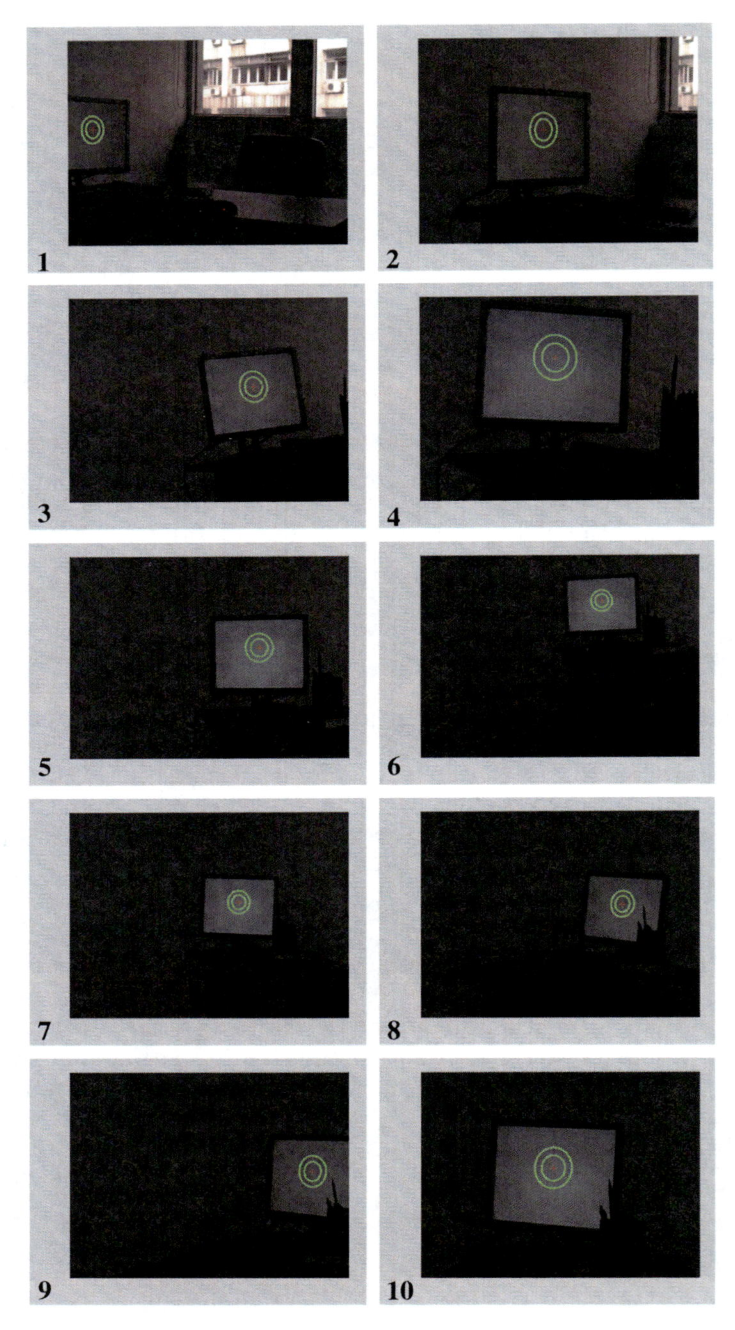

Fig. 3.13 TEN images captured during the travel of the vision system in our lab, labeled by 1, 2,. . . ,10 from left to right and top to down. The RED crosses illustrate the detected positions of the imaged circle center

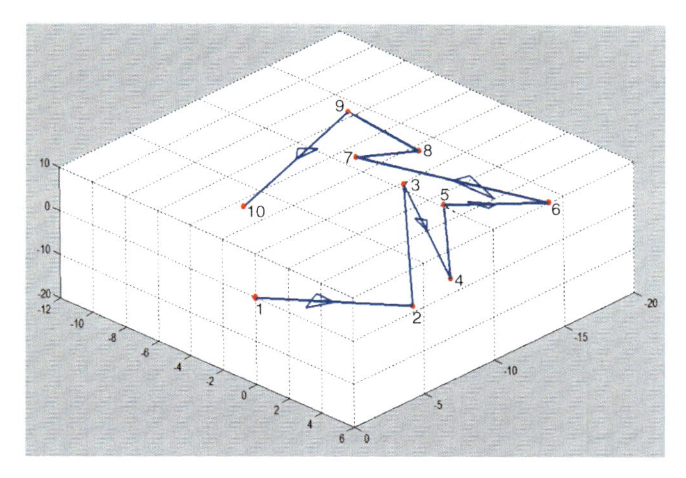

Fig. 3.14 The reconstructed motion trajectory in the experiment, in which the numbers of 1, 2,. . . , 10 represent different positions of the vision system

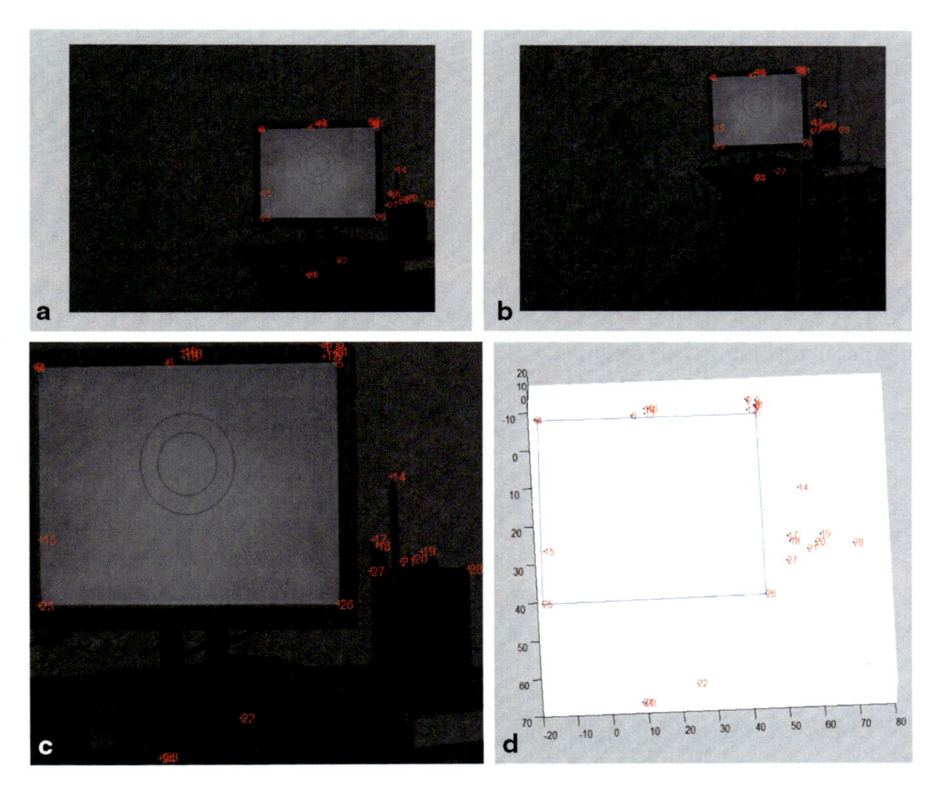

Fig. 3.15 Results of the 3D reconstruction. **a** and **b** show the two images together with extracted feature points plotted by RED ' + '. **c** gives a close look at the fifth image. **d** shows the 28 reconstructed 3D points

Table 3.1 The true and estimation angles subtended by three reconstructed point (unit: °)

Related corner points	4th, 25th and 26th points	25th, 26th and 5th points	26th, 5th and 4th points	5th, 4th and 25th points
True values	90	90	90	90
Estimation values	88.6361	89.6050	90.6788	91.0695

Figure 3.15d provides all the reconstructed 3D points. It is noted that the 4th, 5th, 25th and 26th neighboring points are connected to obtain four line segments, which represent the inner edges of the computer screen. We can see that the relative positions of those points accord with the real scene. The parallelism and perpendicularity between the line segments are well recovered. For example, the angles subtended by the four corner points of the computer screen are given in Table 3.1, in which it is observed that the estimated values of those angles are very close to their true values.

3.5 Line-Based Distortion Correction

In the above pinhole model, we assume that the optical center of the camera, the world point and its image pixel and are collinear, and world line imaged as straight line, and so on. In practice, this assumption does not hold especially for larger view-field of the lens, which means there is some nonlinear distortion in the image. Figure 3.16 shows a comparison, in which the left gives the image of a rectangle under pinhole model while the right under barrel distortion model. We can see that the Barrel distortion introduces nonlinear changes in the image, due to which, image areas near the distortion center are compressed less, while areas farther from the center are compressed more. Because of this, the outer areas of the image look significantly smaller than their actual size.

The nonlinear distortion parameters as well as the focal length and principal point are internal parameters of a camera. It is an important problem for accurate calibration in photo-grammetry, digital image analysis and computer vision. And a variety of methods have been proposed to calibrate different types of cameras with lens distortion in the past decades. These methods fall into two classes: the total calibration algorithm [180, 181] and the nonmetric calibration algorithm [182, 183]. The total calibration methods use a calibration object whose geometrical structure is accurately known. In this class of methods, distortion parameters are obtained together with other intrinsic and extrinsic parameters of the camera. Due to coupling between distortion parameters and other intrinsic as well as extrinsic parameters, high errors of calibration may be resulted in. The nonmetric methods utilize projecting invariants, not relying on calibration object of known structure. In this class of methods, the most important invariant is straight line, and straight-line based methods [182, 183] are the mainstream in calibration of lens distortion so far.

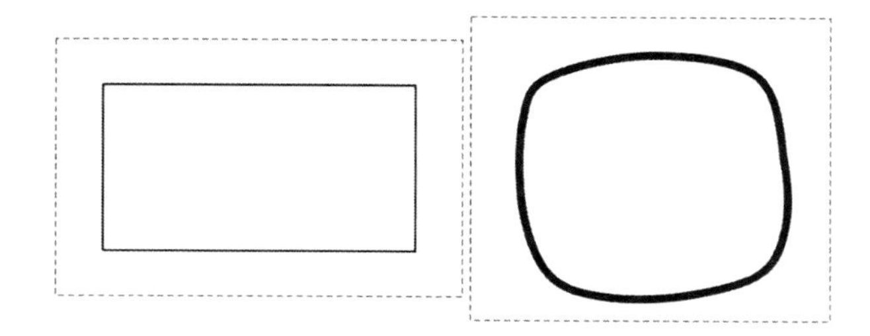

Fig. 3.16 Images taken by pinhole camera model and real camera

3.5.1 The Distortion Model

In order to correct the nonlinear distortions, a suitable distortion model is needed. In the literatures, there are various distortion models for the camera lens, such as Division model, Bicubic model and rational function model, etc. In this section, we will introduce a full-parameter model for our camera [184].

Let $\boldsymbol{m} = [u \quad v]^{\mathrm{T}}$ be an image point in real camera and $\overline{\boldsymbol{m}} = [\overline{u} \quad \overline{v}]^{\mathrm{T}}$ in ideal pinhole camera. Their relationship can be described by

$$\begin{cases} \overline{u} = u + \delta_u(u, v) \\ \overline{v} = v + \delta_v(u, v) \end{cases} \tag{3.19}$$

In (3.19), δ_u and δ_v are functions of the variables u and v, representing the non-linear distortion items in U-axis and V-axis of the image respectively. Mathematically, they can be formulated as

$$\begin{cases} \delta_u = k_1 u(u^2 + v^2) + (p_1(3u^2 + v^2) + 2p_2 uv) + s_1(u^2 + v^2) \\ \delta_v = k_2 v(u^2 + v^2) + (p_2(3u^2 + v^2) + 2p_1 uv) + s_2(u^2 + v^2) \end{cases} \tag{3.20}$$

where k_1, k_2 denote the radial distortion, p_1, p_2 denote the decentering distortion and s_1, s_2 denote the thin prism distortion.

So there are totally six nonlinear distortion parameters within this model. In general, they have very small values. In the extreme case, this model becomes the classical pinhole model if all the nonlinear parameters equal to zeros.

From (3.20), we find that the closer the image points to the image center, the less the distortion effects since the values of the elements u and v are smaller. The effect can be ignored in the center area. On the contrary, the distortion becomes severe for pixels close to the edge of the image. This fact is illustrated in Fig. 3.17.

Based on the above observations, we can make a bit modification on the previous planar pattern for the distortion correction as shown in Fig. 3.18. In this pattern, each edge of the equilateral dodecagon is extended, where the center part is used to approximate the line and other parts for estimating the distortion parameters.

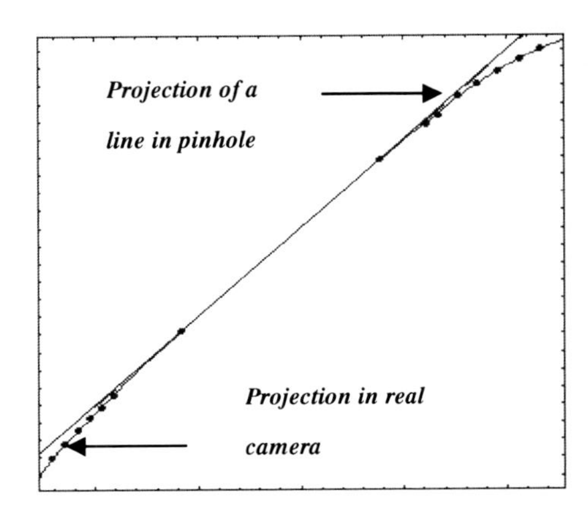

Fig. 3.17 Comparison of projections under the two models

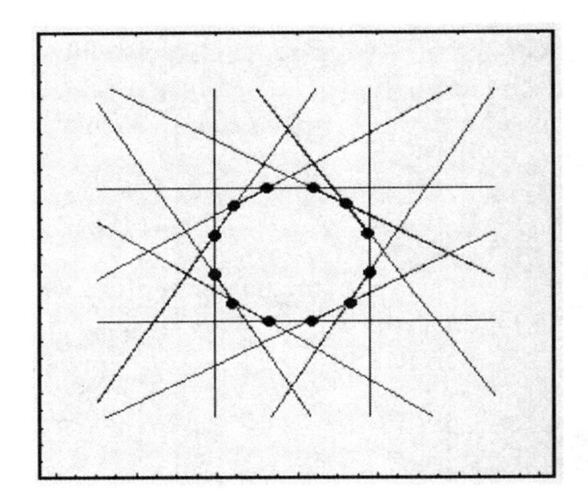

Fig. 3.18 The suggested planar pattern

3.5.2 *The Correction Procedure*

On the whole, the distortion correction process using the suggested pattern can be divided into two stages. One is to compute the lines in the image and the other is to estimate the distortion parameters based on the distortion model. The procedure can be summarized as the following four steps.

1. For each edge of the polygon, extracting several projection points between the two vertexes and estimating the line vector denoted by $\boldsymbol{l}_j = [a_j \quad b_j \quad c_j]^{\mathrm{T}}$.
2. Extracting some points on both outer sides, denoted by $\boldsymbol{m}_i = [u_i \quad v_i]^{\mathrm{T}}$. They should lie on $\boldsymbol{l}_j$ in the pinhole model, i.e. $\boldsymbol{l}_j^{\mathrm{T}}\boldsymbol{m}_i = 0$. So we have the constraints on the distortion parameters $f_i(k_1, k_2, p_1, p_2, s_1, s_2) = a_j(u_i + \delta_u(u_i, v_i)) + b_j(v_i + \delta_v(u_i, v_i)) + c_j = 0$
3. Repeating the above two steps, we obtain a group of equations $F_{ij}(k_1, k_2, p_1, p_2, s_1, s_2) = 0$

where j denotes the sequence of line and i denotes the corresponding point sequence.

4. Solving the equations by nonlinear optimization, we can achieve the values for $k_1, k_2, p_1, p_2, s_1, s_2$.

Remark In this procedure, it is very important to extract the projection points evenly, both for calculating the line vector and for estimating the distortion parameters. In other words, those data we used should be representative. Besides, a good optimization algorithm should be employed since the computational complexity is expensive when more points are involved. In our experiment, the classical Levenberg–Marquardt algorithm shows a good performance. Since only the image of lines is employed, this technique is named as line-based distortion correction.

3.5.3 Examples

The planar pattern given in Fig. 3.18 is used for the experiments of static calibration, including calibration of the intrinsic parameters and the distortion correction. Once those parameters have been obtained, we can test on the real images. Figure 3.19 shows the effect of distortion correction with our camera. The left is the original image of the wall from a building in our university, in which the lines of the outer part bear obvious distortions. And the right is the corrected image. We can see that the lines become lines and the quality of the image has been improved significantly. In another experiment, the image of the whole building is captured and given in Fig. 3.20a. Figure 3.20b shows the corrected image. It is noticed that improvement in the image is obtained, especially around the outside contour of the building.

3.6 Summary

The camera calibration has always been a classical and critical problem in the field of computer vision. In this chapter, we have designed three different planar patterns for performing this task. The first pattern contains an equilateral polygon, whose parallel edges are used to estimate the vanishing line and vertexes for the ellipse in the camera image. Then the image of circular point and hence the camera parameters can be computed. Two intersecting circles are involved in the second pattern and the problem becomes the solution of several quartic equations with one variable. In the third pattern, a pair of concentric circles are employed, in which the constraints on the projection of the circle center are firstly established by considering the pole-polar relationship and then formulating the solution into a first-order polynomial eigenvalue problem. There are some characteristics for these patterns: Firstly, it is easy to prepare the planar patterns since the positions and sizes and the concerned entities can be arbitrary; Secondly, the work of image processing is simple as no

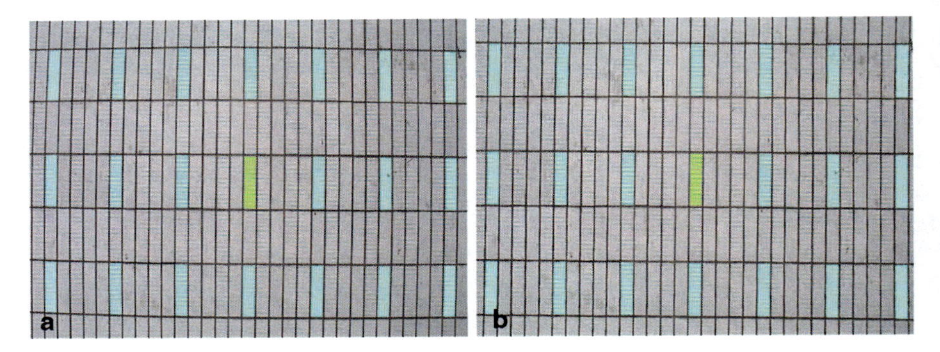

Fig. 3.19 An experiment on the wall of a building

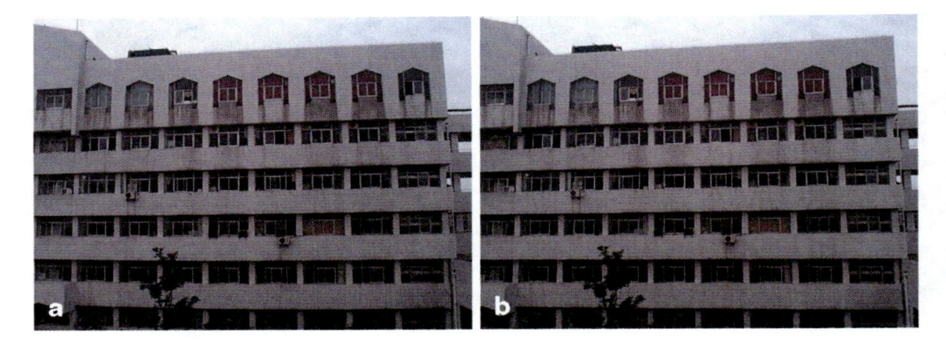

Fig. 3.20 Another test on a building

complicated structure or feature matching are needed in the algorithms. Lastly, the calibration algorithm is simple to implement and the computation is efficient. Some numerical simulations and real data experimental results are presented to show their validity and efficiency.

Another work in this chapter is to estimate the distortion parameters of a camera with a line-based method. This is based on the observation of the image and the mathematical formula of distortion models. The examples of distortion correction given in Sect. 3.5.3 demonstrate obvious improvements on the quality of the images. We find that this process is computationally intensive, especially when full distortion model is used. In practice, a lookup table can be setup for distortion correction. This will greatly improve the efficiency since those distortion parameters are intrinsic for each camera.

Chapter 4
Homography-Based Dynamic Calibration

The relative pose problem is to find the possible solutions for relative camera pose between two calibrated views given several corresponding coplanar or non-coplanar points. In this chapter, we will talk about an algorithm for determining the relative pose in a structured light system using coplanar points, supposing that its intrinsic parameters have been obtained by static calibration as discussed in the previous chapter. The image-to-image Homographic matrix is extensively explored with the assumption of one arbitrary plane in the scene. A closed-form solution is provided to obtain computational efficiency. Redundancy in the data is easily incorporated to improve the reliability of the estimations in the presence of noise. By using the matrix perturbation theory, we give a typical sensitivity analysis on the estimated pose parameters with respect to the noise in the image points. Finally, some experimental results are shown to valid this technique.

4.1 Problem Statement

In a dynamic environment, the objects may be different in sizes and distances, and the task requirements may also be different for different applications. In such cases, a structure-fixed vision system does not work well. On the other hand, a reconfigurable system can change its structural parameters to adapt itself to the scene to obtain maximum 3D information. Whenever the reconfiguration occurs, the relative motion problem arises, which deals with finding the relative orientation and position in a vision system so that the 3D reconstruction can be followed immediately. This is also named as the relative pose problem. A good solution for it will greatly improve the efficiency of a vision system. In this chapter, we will study this problem in the context of a structured light stereovision system, which consists of a DLP projector and a CCD camera.

In practice, various types of errors occur during the process of calibration, which will lead to errors in the calibration results and affect the system's performance. Some errors, such as optical distortion errors, are stable and repeatable, and thus easy to predict and correct them via the mathematical models, just as what we have done in

B. Zhang, Y. F. Li, *Automatic Calibration and Reconstruction for Active Vision Systems*, Intelligent Systems, Control and Automation: Science and Engineering, DOI 10.1007/978-94-007-2654-3_4, © Springer Science+Business Media B.V. 2012

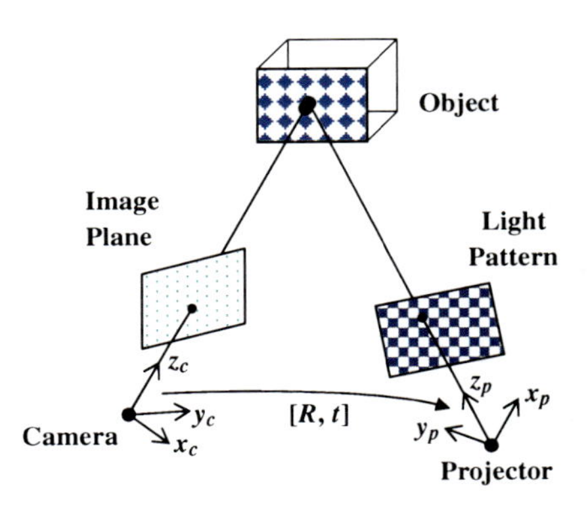

Fig. 4.1 Geometrical relations of the vision system

Chap. 3. Some errors, such as feature detector errors and matching errors, vary from time to time and thus require special efforts to deal with. Due to the importance of sensitivity analysis, much work has been conducted in the relevant research. In [166], Criminisi et al. developed an uncertainty analysis method including both the errors in image localization and the uncertainty in the imaging transformation for their plane measuring device. Grossmann et al. [167] and Sun et al. [168] independently presented their work on the uncertainty analysis for 3D reconstruction. Recently, Wang [169] proposed an error analysis of the intrinsic camera parameters for pure rotation-based calibration in the presence of small translation. In these methods, the error sensitivity is analyzed via specific algorithms. As a result, they are not applicable for the relative pose problem in our structured light system. Therefore, another task in this work is to analyze the error sensitivity of the dynamic calibration.

Here, the projector is considered as a dual camera, which projects rather than receives the light rays. So the structured light system can be used as an active stereoscopic vision system. For the camera and the projector, we define a right-handed coordinate system originated at their optical centers respectively (see Fig. 4.1). Let $\boldsymbol{R}$ and $\boldsymbol{t}$ denote the rotation matrix and translation vector from the camera to the projector. We assume that the world coordinate system coincides with that of the camera. Then, for an arbitrary 3D point $\boldsymbol{M}$, its image projections in the camera and the projector can be expressed as

$$\boldsymbol{m}_c = \alpha \boldsymbol{K}_c \boldsymbol{M} \tag{4.1}$$

and

$$\boldsymbol{m}_p = \sigma \boldsymbol{K}_p (\boldsymbol{R}\boldsymbol{M} + \boldsymbol{t}) \tag{4.2}$$

where $\boldsymbol{m}_c = [u \quad v \quad 1]^{\mathrm{T}}$ and $\boldsymbol{m}_p = [u' \quad v' \quad 1]^{\mathrm{T}}$ are the projection points on the image plane and the projector plane, α and σ are nonzero scale factors. Next, we show how to recover $\boldsymbol{M}$ through (4.1) and (4.2).

Supposing $\boldsymbol{K}_c$ and $\boldsymbol{K}_p$ are defined in Chap. 2. Let $\boldsymbol{K}_p\boldsymbol{R} = \begin{bmatrix} \boldsymbol{k}_1 \\ \boldsymbol{k}_2 \\ \boldsymbol{k}_3 \end{bmatrix}$ and $\boldsymbol{K}_p\boldsymbol{t} = \begin{bmatrix} k_1 \\ k_2 \\ k_3 \end{bmatrix}$. From (4.1) and (4.2), we have four equations on the coordinates of point $\boldsymbol{M}$, written in matrix form as

$$\boldsymbol{X}\boldsymbol{M} = \boldsymbol{x} \tag{4.3}$$

where $\boldsymbol{X} = \begin{bmatrix} & u'\boldsymbol{k}_3 - \boldsymbol{k}_1 & \\ & v'\boldsymbol{k}_3 - \boldsymbol{k}_2 & \\ f_u & s & u_0 - u \\ 0 & f_v & v_0 - v \end{bmatrix}$ and $\boldsymbol{x} = \begin{bmatrix} k_1 - k_3 u' \\ k_2 - k_3 v' \\ 0 \\ 0 \end{bmatrix}$.

In the least squares sense, the 3D world point can be determined by

$$\boldsymbol{M} = \left(\boldsymbol{X}^{\mathrm{T}}\boldsymbol{X}\right)^{-1}\boldsymbol{X}^{\mathrm{T}}\boldsymbol{x} \tag{4.4}$$

This formula describes the basic principle for 3D reconstruction using traditional triangulation method. To obtain an optimal result, this method can be used as an initial estimation for the nonlinear optimization problem with the Levenberg-Marquardt algorithm.

From (4.3), once the intrinsic and extrinsic parameters of the camera and the projector are obtained, we can compute the matrix $\boldsymbol{X}$ and the vector $\boldsymbol{x}$. Then the triangulation method for computing the 3-D coordinates of a point simply involves calculating (4.4) given its correspondent projections. Thus the whole calibration of the structured light system can be divided into two parts. The first part concerns the calibration of intrinsic parameters (such as focal lengths and principle point) of the camera and the projector, called static calibration. The static calibration needs to be performed only once. We assume that it has been done as in Chap. 3. The second part deals with calibration of the extrinsic parameters of the relative pose, called dynamic calibration. This process should be repeated whenever the system changes its configuration. There are 6 unknown extrinsic parameters since arbitrary motion is considered, i.e. three for 3-axis rotation and three for 3-dimensional translation. The dynamic calibration task is to determine these parameters and analyze their error sensitivity, which is the focus of this chapter.

4.2 System Constraints

4.2.1 Two Propositions

Before starting the details of this technique, we establish two propositions which should be used for the following dynamic calibration.

Proposition 4.1 *Let* $\boldsymbol{g}$ *be any* 3×1 *nonzero vector and* $\boldsymbol{G}$ *be a* 3×3 *nonzero symmetric matrix. If* $[\boldsymbol{g}]_\times \boldsymbol{G}[\boldsymbol{g}]_\times = \boldsymbol{0}$*, then the determinant of* $\boldsymbol{G}$ *is zero.*

Proof

Let $\boldsymbol{g} = [g_1 \quad g_2 \quad g_3]^{\mathrm{T}}$ and $\boldsymbol{G} = \begin{bmatrix} G_{11} & G_{12} & G_{13} \\ G_{12} & G_{22} & G_{23} \\ G_{13} & G_{23} & G_{33} \end{bmatrix}$

From $[\boldsymbol{g}]_\times \boldsymbol{G}[\boldsymbol{g}]_\times = 0$, we will have six equations shown as follows

$$\begin{cases} g_2^2 G_{11} - 2g_1 g_2 G_{12} + g_1^2 G_{22} = 0 \\ g_3^2 G_{22} - 2g_2 g_3 G_{23} + g_2^2 G_{33} = 0 \\ g_3^2 G_{11} - 2g_1 g_3 G_{13} + g_1^2 G_{33} = 0 \\ g_3^2 G_{12} - g_1 g_3 G_{23} - g_2 g_3 G_{13} + g_1 g_2 G_{33} = 0 \\ g_2^2 G_{13} - g_2 g_3 G_{12} + g_1 g_3 G_{22} - g_1 g_2 G_{23} = 0 \\ g_1^2 G_{23} + g_2 g_3 G_{11} - g_1 g_2 G_{13} - g_1 g_3 G_{12} = 0 \end{cases} \tag{4.5}$$

We take three cases to prove this proposition.

Case 1: none of the elements in the vector $\boldsymbol{g}$ is zero

From the first equation in (4.5), we have

$$G_{11} = g_1 \left(\frac{2g_2 G_{12} - g_1 G_{22}}{g_2^2} \right) \tag{4.6}$$

Let $f_1 = \frac{2g_2 G_{12} - g_1 G_{22}}{2g_2^2}$.

Similarly, from the second and third equations in (4.5), we can have $f_2 = \frac{2g_3 G_{23} - g_2 G_{33}}{2g_3^2}$ and $f_3 = \frac{2g_1 G_{13} - g_3 G_{11}}{2g_1^2}$.

Let the vector $\boldsymbol{f} = [f_1 \quad f_2 \quad f_3]^{\mathrm{T}}$. It is easy to verify that $\boldsymbol{G} = \boldsymbol{f}\boldsymbol{g}^{\mathrm{T}} + \boldsymbol{g}\boldsymbol{f}^{\mathrm{T}}$ from (4.5). Hence, the matrix $\boldsymbol{G}$ is rank deficiency and $\det(\boldsymbol{G}) = 0$.

Case 2: only one of the elements in the vector $\boldsymbol{g}$ is zero

Without lost of generality, we assuming that $g_1 = 0$, $g_2 \neq 0$ and $g_3 \neq 0$. From the first equation in (4.5), we have $G_{11} = 0$. From the fourth equation in (4.5), we have

$$g_3 = \frac{g_2 G_{13}}{G_{12}} \tag{4.7}$$

Substituting (4.7) into the second equation in (4.5), we get

$$G_{12}^2 G_{33} + 2G_{12}G_{13}G_{23} - G_{13}^2 G_{22} = 0 \tag{4.8}$$

This is exactly the determinant of $\boldsymbol{G}$.

Case 3: two of the elements in the vector $\boldsymbol{g}$ are zero

Without lost of generality, we assuming that $g_1 = g_2 = 0$ and $g_3 \neq 0$. From the second, third and fourth equations in (4.5), it is easy to verify that $G_{11} = G_{12} = G_{22} = 0$, from which we have $\det(\boldsymbol{G}) = 0$.

In summary, the determinant of $\boldsymbol{G}$ is zero. □

Proposition 4.2 *Let $\boldsymbol{f}$ and $\boldsymbol{g}$ be any two 3×1 non-zero vectors. Assuming the three eigenvalues of the matrix $\boldsymbol{I} + \boldsymbol{fg}^{\mathrm{T}} + \boldsymbol{gf}^{\mathrm{T}}$ are δ_1, δ_2 and δ_3, then either $\delta_1 < 1, \delta_2 = 1, \delta_3 > 1$ or $\delta_1 < 1, \delta_2 = \delta_3 = 1$.*

Proof Let $\boldsymbol{f} = [f_1 \;\; f_2 \;\; f_3]^{\mathrm{T}}$ and $\boldsymbol{g} = [g_1 \;\; g_2 \;\; g_3]^{\mathrm{T}}$.

From the definition of characteristic function, we have

$$\det\left(\boldsymbol{I} + \boldsymbol{fg}^{\mathrm{T}} + \boldsymbol{gf}^{\mathrm{T}} - \delta \boldsymbol{I}\right) = 0 \tag{4.9}$$

Expanding (4.9), we obtain

$$(1 - \delta)\left((1 - \delta)^2 + p(1 - \delta) + q\right) = 0 \tag{4.10}$$

where $p = -2(f_1 g_1 + f_2 g_2 + f_3 g_3)$ and

$$q = -(f_1 g_2 - f_2 g_1)^2 - (f_1 g_3 - f_3 g_1)^2 - (f_2 g_3 - f_3 g_2)^2.$$

From (4.10), we can see that $\delta_2 = 1$ always is a solution.

Let

$$F(\delta) = (1 - \delta)^2 + p(1 - \delta) + q = 0 \tag{4.11}$$

We take three cases to investigate the solutions of $F(\delta)$.

Case 1: $q = 0$

When $q = 0$, we have $f_1/g_1 = f_2/g_2 = f_3/g_3$, which indicates that vector $\boldsymbol{f}$ is parallel with $\boldsymbol{g}$. So we have $p < 0$.

From (4.11), we have $(1 - \delta)(1 + p - \delta) = 0$. Hence, the solutions of $F(\delta)$ are $\delta_1 < 1$ and $\delta_3 = 1$.

Case 2: $p = 0$

It is obvious that $p = 0$ means that $\boldsymbol{f}$ is orthogonal to $\boldsymbol{g}$. Since $\boldsymbol{f}$ cannot be parallel with $\boldsymbol{g}$ at the same time, q can not be zero. From (4.11), the solutions of $F(\delta)$ are $\delta_1 < 1$ and $\delta_3 > 1$.

Case 3: $q \neq 0$ and $p \neq 0$

From $\Delta = p^2 - 4q$, we have

$$\Delta = 4(f_1 g_1 + f_2 g_2 + f_3 g_3)^2 + 4(f_1 g_2 - f_2 g_1)^2 + 4(f_1 g_3 - f_3 g_1)^2 + 4(f_2 g_3 - f_3 g_2)^2.$$

Again, since $\boldsymbol{f}$ cannot be both orthogonal to and parallel with $\boldsymbol{g}$ at the same time, we have $\Delta \neq 0$. Therefore, the solutions of $F(\delta)$ is $\delta_1 < 1$ and $\delta_3 > 1$ in this case.

In summary, the three eigenvalues of matrix $\boldsymbol{I} + \boldsymbol{f}\boldsymbol{g}^{\mathrm{T}} + \boldsymbol{g}\boldsymbol{f}^{\mathrm{T}}$ are either $\delta_1 < 1$, $\delta_2 = 1, \delta_3 > 1$ or $\delta_1 < 1, \delta_2 = \delta_3 = 1$. □

4.2.2 System Constraints

Assuming the equation of the planar structure is $\boldsymbol{n}^{\mathrm{T}}\boldsymbol{M} = 1$, here $\boldsymbol{n}$ is the normal vector of the plane, from (4.2), we have

$$\boldsymbol{m}_p = \sigma \boldsymbol{K}_p \left(\boldsymbol{R} + \boldsymbol{t}\boldsymbol{n}^{\mathrm{T}}\right) \boldsymbol{M} \tag{4.12}$$

Combining (4.1) and (4.12),

$$\boldsymbol{m}_p = \frac{\sigma}{\alpha} \boldsymbol{K}_p \left(\boldsymbol{R} + \boldsymbol{t}\boldsymbol{n}^{\mathrm{T}}\right) \boldsymbol{K}_c^{-1} \boldsymbol{m}_c \tag{4.13}$$

Considering (2.9) and (4.13), the explicit formula for the plane-based Homography is

$$\lambda \boldsymbol{H} = \boldsymbol{K}_p \left(\boldsymbol{R} + \boldsymbol{t}\boldsymbol{n}^{\mathrm{T}}\right) \boldsymbol{K}_c^{-1} \tag{4.14}$$

where λ is a scale factor.

The equivalent form of (4.14) is

$$\lambda \overline{\boldsymbol{H}} = \lambda \boldsymbol{K}_p^{-1} \boldsymbol{H} \boldsymbol{K}_c = \boldsymbol{R} + \boldsymbol{t}\boldsymbol{n}^{\mathrm{T}} \tag{4.15}$$

where $\overline{\boldsymbol{H}}$ is called the calibrated Homography.

Since $\boldsymbol{K}_p$ and $\boldsymbol{K}_c$ are calibrated and the Homographic matrix $\boldsymbol{H}$ can be computed from (2.10) using corresponding points, $\overline{\boldsymbol{H}}$ is known.

Multiplying matrix $[\boldsymbol{t}]_\times$ on both sides of (4.15), we have

$$\lambda [\boldsymbol{t}]_\times \overline{\boldsymbol{H}} = [\boldsymbol{t}]_\times \boldsymbol{R} \tag{4.16}$$

It is well known that the right side of (4.16) is named as the Essential matrix. So this equation reveals the relationship between the calibrated Homographic matrix and the Essential matrix.

As $\boldsymbol{R}$ is a rotation matrix, $\boldsymbol{R}\boldsymbol{R}^{\mathrm{T}} = \boldsymbol{I}$. From (4.16), we obtain

$$\lambda^2 [\boldsymbol{t}]_\times \overline{\boldsymbol{H}\boldsymbol{H}}^{\mathrm{T}} [\boldsymbol{t}]_\times = [\boldsymbol{t}]_\times [\boldsymbol{t}]_\times \tag{4.17}$$

Rearranging (4.17) gives

$$[\boldsymbol{t}]_\times \boldsymbol{W} [\boldsymbol{t}]_\times = \boldsymbol{0} \tag{4.18}$$

where $\boldsymbol{W} = \lambda^2 \overline{\boldsymbol{H}\boldsymbol{H}^{\mathrm{T}}} - \boldsymbol{I}$.

The above formula provides constraints on the translation vector from the Homographic matrix. In the following section, we will tell how to calibrate the relative pose parameters based on these constraints.

4.3 Calibration Algorithm

4.3.1 Solution for the Scale Factor

In (4.18), it is easy to prove that the matrix $\boldsymbol{W}$ is symmetric by $\boldsymbol{W}^{\mathrm{T}} = \boldsymbol{W}$. Since $\boldsymbol{t}$ represents the translation vector between the camera and the projector, it cannot be zero vector.

According to proposition 4.1, we have

$$\det(\boldsymbol{W}) = \det\left(\lambda^2 \overline{\boldsymbol{H}\boldsymbol{H}}^{\mathrm{T}} - \boldsymbol{I}\right) = 0 \tag{4.19}$$

(4.19) indicates that λ^2 is the inverse of eigenvalues of matrix $\overline{\boldsymbol{H}\boldsymbol{H}}^{\mathrm{T}}$.

There are three eigenvalues for $\overline{\boldsymbol{H}\boldsymbol{H}}^{\mathrm{T}}$. Next, we will discuss which one is what we need by proposition 4.2.

From (4.15), we have

$$\lambda^2 \overline{\boldsymbol{H}\boldsymbol{H}}^{\mathrm{T}} = \left(\boldsymbol{R} + \boldsymbol{t}\boldsymbol{n}^{\mathrm{T}}\right) * \left(\boldsymbol{R} + \boldsymbol{t}\boldsymbol{n}^{\mathrm{T}}\right)^{\mathrm{T}} \tag{4.20}$$

This can be equivalently changed into

$$\begin{aligned}
\lambda^2 \overline{\boldsymbol{H}\boldsymbol{H}}^{\mathrm{T}} &= \boldsymbol{I} + \boldsymbol{R}\boldsymbol{t}\boldsymbol{n}^{\mathrm{T}} + \boldsymbol{t}\boldsymbol{n}^{\mathrm{T}}\boldsymbol{R}^{\mathrm{T}} + \boldsymbol{n}^{\mathrm{T}}\boldsymbol{n}\boldsymbol{t}\boldsymbol{t}^{\mathrm{T}} \\
&= \boldsymbol{I} + \left(\boldsymbol{R}\boldsymbol{n} + \frac{\boldsymbol{n}^{\mathrm{T}}\boldsymbol{n}}{2}\boldsymbol{t}\right)\boldsymbol{t}^{\mathrm{T}} + \boldsymbol{t}\left(\boldsymbol{n}^{\mathrm{T}}\boldsymbol{R}^{\mathrm{T}} + \frac{\boldsymbol{n}^{\mathrm{T}}\boldsymbol{n}}{2}\boldsymbol{t}^{\mathrm{T}}\right) \\
&= \boldsymbol{I} + \boldsymbol{s}\boldsymbol{t}^{\mathrm{T}} + \boldsymbol{t}\boldsymbol{s}^{\mathrm{T}}
\end{aligned} \tag{4.21}$$

where $\boldsymbol{s} = \boldsymbol{R}\boldsymbol{n} + \frac{\boldsymbol{n}^{\mathrm{T}}\boldsymbol{n}}{2}\boldsymbol{t}$.

Without lose of generality, we assume that the scene plane is opaque. Hence, the camera and the projector should lie on the same side of the scene plane and locate at different positions. As a result, neither $\boldsymbol{t}$ nor $\boldsymbol{s}$ is zero vector. According to proposition 4.2, $\lambda^2 \overline{\boldsymbol{H}\boldsymbol{H}}^{\mathrm{T}}$ or $\boldsymbol{I} + \boldsymbol{s}\boldsymbol{t}^{\mathrm{T}} + \boldsymbol{t}\boldsymbol{s}^{\mathrm{T}}$, will have one eigenvalue as 1, which lies between the other two different eigenvalues or which is as the eigenvalue with multiplicity two. Since the eigenvalues of $\overline{HH}^T$ are $1/\lambda^2$ times of those of $\lambda^2\overline{HH}^T$, we have the following conclusions:

1. If the three eigenvalues are distinct from each other, $1/\lambda^2$ is the eigenvalue of $\overline{HH}^T$ that lies between the other two;
2. If one of the eigenvalues of $\overline{HH}^T$ is of multiplicity two, $1/\lambda^2$ is equal to this eigenvalue.

By these conclusions, we can obtain the solution for λ^2 that satisfies (4.17). So is the value for λ, whose sign can be determined by considering the consistency of both sides in (2.9) for each pair of points.

4.3.2 Solutions for the Translation Vector

After λ^2 is obtained, the matrix $\boldsymbol{W}$ in (4.18) is known. Then the constraints on the translation vector can be reformulated into the following six equations:

$$w_{33}{t_1}^2 - 2w_{13}t_1t_3 + w_{11}{t_3}^2 = 0 \quad (4.22.1)$$

$$w_{33}{t_2}^2 - 2w_{23}t_2t_3 + w_{22}{t_3}^2 = 0 \quad (4.22.2)$$

$$w_{33}t_1t_2 - w_{23}t_1t_3 - w_{13}t_2t_3 + w_{12}{t_3}^2 = 0 \quad (4.22.3)$$

$$w_{13}{t_2}^2 - w_{23}t_1t_2 + w_{22}t_1t_3 - w_{12}t_2t_3 = 0 \quad (4.22.4)$$

$$w_{23}{t_1}^2 - w_{13}t_1t_2 - w_{12}t_1t_3 + w_{11}t_2t_3 = 0 \quad (4.22.5)$$

$$w_{22}{t_1}^2 + w_{11}{t_2}^2 - 2w_{12}t_1t_2 = 0 \quad (4.22.6)$$

where w_{ij} denotes the ij-th element of matrix $\boldsymbol{W}$.

Here, we will talk about three different approaches for solving the translation vector from these equations.

I. Solution from an Algebraic Viewpoint (4.22) provides six constraints on the translation vector, which can be used to give the solutions algebraically. Here, the translation vector is determined up to a nonzero scale factor. For simplicity, we assume its third component is unit, i.e. $t_3 = 1$.

Clearly, t_1, t_2 can be obtained analytically from the first two equations, which should satisfy the last four equations in general. In case of noise data, these six equations are used to solve the translation vector in the least squares sense.

II. Solution from a Geometric Viewpoint Using polynomial elimination method, we can obtain the following three linear equations from the six quadratic equations in (4.22) (Please refer to Appendix A for more details):

$$\begin{cases} a_1 t_1 + b_1 t_2 + c_1 = 0 \\ a_2 t_1 + b_2 t_2 + c_2 = 0 \\ a_3 t_1 + b_3 t_2 + c_3 = 0 \end{cases} \tag{4.23}$$

where $a_1 = w_{13}w_{23} - w_{12}w_{33}, \quad b_1 = w_{11}w_{33} - w_{13}^2, \quad c_1 = w_{12}w_{13} - w_{11}w_{23}$

$a_2 = w_{22}w_{33} - w_{23}^2, \quad b_2 = w_{13}w_{23} - w_{12}w_{33}, \quad c_2 = w_{12}w_{23} - w_{13}w_{22}$

$a_3 = w_{22}w_{13} - w_{12}w_{23}, \quad b_3 = w_{11}w_{23} - w_{12}w_{13}, \quad c_3 = w_{12}^2 - w_{11}w_{22}.$

Proposition 4.3 *The three equations in* (4.23) *are equivalent to each other.*

Proof Considering (4.21) and the definition of $\boldsymbol{W}$ in (4.18), we have $\boldsymbol{W} = \boldsymbol{s}\boldsymbol{t}^{\mathrm{T}} + \boldsymbol{t}\boldsymbol{s}^{\mathrm{T}}$. Hence, the rank of $\boldsymbol{W}$ is deficiency, and its determinant should be zero, i.e.

$$\det(\boldsymbol{W}) = 2w_{12}w_{13}w_{23} - w_{23}^2 w_{11} - w_{12}^2 w_{33} - w_{13}^2 w_{22} + w_{11}w_{22}w_{33} = 0$$

With some algorithmic operations, we have

$$a_1 b_2 - a_2 b_1 = w_{33} \det(\boldsymbol{W}) = 0$$

and

$$c_1 b_2 - c_2 b_1 = w_{13} \det(\boldsymbol{W}) = 0.$$

Therefore, the first two equations in (4.23) are equivalent to each other.

In a similar way, the first and the third equations in (4.23) can be proved to be equivalent to each other as shown by:

$$a_1 b_3 - a_3 b_1 = w_{13} \det(\boldsymbol{W}) = 0$$

and

$$c_1 b_3 - c_3 b_1 = w_{11} \det(\boldsymbol{W}) = 0.$$

In summary, the three equations in (4.23) are equivalent to each other. □

According to this proposition, one of the three linear equations and one of the six quadratic equations can be chosen to get an analytic solution for the translation vector. Geometrically, the solution for the translation vector is equivalent to the intersection points of a line in (4.23) and a conic in (4.22). Here, the solution is given from the first equation in (4.22) and the first one in (4.23):

$$\begin{cases} t_1 = \dfrac{w_{13} \pm \sqrt{w_{13}^2 - w_{11}w_{33}}}{w_{33}} \\ t_2 = \dfrac{(w_{13}w_{23} - w_{12}w_{33})t_1 + (w_{12}w_{13} - w_{11}w_{23})}{w_{11}w_{33} - w_{13}^2} \\ t_3 = 1 \end{cases} \tag{4.24}$$

III. Homogeneous Solution In the above two cases, an inhomogeneous solution is obtained, where t_3 has been chosen to be 1 or any other nonzero value. However, the solution will be unreliable when it is really zero or close to zero. To overcome this drawback, a homogeneous solution is proposed.

We first define a 6-D vector $\boldsymbol{\tau} = \begin{bmatrix} t_1^2 & t_2^2 & t_3^2 & t_1t_2 & t_2t_3 & t_2t_3 \end{bmatrix}^{\mathrm{T}}$.

From (4.22), we have

$$\boldsymbol{K}\boldsymbol{\tau} = 0 \tag{4.25}$$

where $\boldsymbol{K} = \begin{bmatrix} w_{33} & 0 & w_{11} & 0 & -2w_{13} & 0 \\ 0 & w_{33} & w_{22} & 0 & 0 & -2w_{23} \\ 0 & 0 & w_{12} & w_{33} & -w_{23} & -w_{13} \\ 0 & w_{13} & 0 & -w_{23} & w_{22} & -w_{12} \\ w_{23} & 0 & 0 & -w_{13} & -w_{12} & w_{11} \\ w_{22} & w_{11} & 0 & -2w_{12} & 0 & 0 \end{bmatrix}$.

Proposition 4.4 *The rank of matrix* $\boldsymbol{K}$ *in* (4.25) *is* 4.

Proof Since the rank of matrix $\boldsymbol{W}$ is 2, the three elements w_{22}, w_{23} and w_{33} in matrix $\boldsymbol{K}$ cannot be zero at the same time, otherwise its rank will be 1. Without lose of generality, we can assume that w_{33} is nonzero. Considering the upper left 4×4 sub-matrix of $\boldsymbol{K}$,

$$\boldsymbol{K}_{44} = \begin{bmatrix} w_{33} & 0 & w_{11} & 0 \\ 0 & w_{33} & w_{22} & 0 \\ 0 & 0 & w_{12} & w_{33} \\ 0 & w_{13} & 0 & -w_{23} \end{bmatrix},$$

the rank of $\boldsymbol{K}_{44}$ is exactly 4 since $\det(\boldsymbol{K}_{44}) = w_{33}^2(w_{13}w_{22} - w_{12}w_{23}) \neq 0$.

Hence, the rank of $\boldsymbol{K}$ is equal to or great than 4.

On the other hand, by iteratively applying row and column operations on $\boldsymbol{K}$, we have

$$\boldsymbol{K} \sim \begin{bmatrix} e_1 & 0 & e_2 & 0 & 0 & 0 \\ 0 & 0 & 0 & e_3 & 0 & 0 \\ 0 & e_4 & e_5 & 0 & 0 & 0 \\ 0 & e_6 & 0 & 0 & 0 & 0 \\ e_7 & 0 & 0 & 0 & 0 & 0 \\ 0 & 0 & 0 & 0 & 0 & 0 \end{bmatrix}$$

where:

$$e_1 = w_{33}, \quad e_2 = \frac{w_{11}(w_{13}w_{23} - w_{12}w_{33})}{w_{13}w_{23}},$$

$$e_3 = \frac{2w_{33}\left(w_{11}w_{12}w_{22}w_{33} - w_{12}w_{13}^2w_{22} + w_{11}w_{13}w_{22}w_{23} - w_{23}^2w_{33}\right)}{w_{13}^2w_{22} - w_{12}^2w_{33}},$$

$$e_4 = -\frac{w_{13}w_{33}}{2w_{23}},\ e_5 = \frac{w_{13}w_{22} - w_{12}w_{23}}{w_{23}},\ e_6 = \frac{w_{12}^2w_{33} + w_{13}^2w_{22} - 2w_{12}w_{13}w_{23}}{2(w_{13}w_{22} - w_{12}w_{23})},$$

$$e_7 = w_{23}.$$

It is obvious that the rank of $\boldsymbol{K}$ is equal to or less than 4 since its last two columns are all zeros.

Therefore, the rank of $\boldsymbol{K}$ must be 4 considering the above two cases. □

According this proposition, we have two eigenvectors whose corresponding singular values are zeroes using singular value decomposition of $\boldsymbol{K}$. Let $\boldsymbol{\tau}_1$ and $\boldsymbol{\tau}_2$ denote the two vectors. Then the solution for vector $\boldsymbol{\tau}$ is $\boldsymbol{\tau}_1 + \boldsymbol{\xi}\boldsymbol{\tau}_2$, here $\boldsymbol{\xi}$ is a scale factor.

From the definition of vector $\boldsymbol{\tau}$, we have

$$\begin{cases} \tau_1\tau_6 - \tau_4\tau_5 = 0 \\ \tau_2\tau_5 - \tau_4\tau_6 = 0 \\ \tau_3\tau_4 - \tau_5\tau_6 = 0 \end{cases} \tag{4.26}$$

where τ_i represents the i-th element in $\boldsymbol{\tau}$.

The ambiguity ξ can be determined from any one of the linear equations in (4.26) or using all in the least squares sense. Then the vector $\boldsymbol{\tau}$ is totally determined.

Now, the homogeneous solution for the translation vector can be obtained from $\boldsymbol{\tau}$ as

$$\begin{cases} t_1 = \pm\sqrt{\tau_1} \\ t_2 = \dfrac{\tau_4}{t_1} \\ t_3 = \dfrac{\tau_5}{t_1} \end{cases} \tag{4.27}$$

4.3.3 Solution for Rotation Matrix

Here, we suggest two different methods for solving this problem:

I. A Direct Solution Since λ, $\boldsymbol{\tau}$ and $\overline{\boldsymbol{H}}$ have been determined in the previous sections, the left side of (4.16), say $\boldsymbol{\Omega} = \lambda[\boldsymbol{t}]_\times\overline{\boldsymbol{H}}$ is known. According to the first and second columns of both sides of (4.16), we have

$$\begin{cases} r_{21} - t_2r_{31} = \Omega_{11} \\ -r_{11} + t_1r_{31} = \Omega_{21} \\ t_2r_{11} - t_1r_{21} = \Omega_{31} \\ r_{11}^2 + r_{21}^2 + r_{31}^2 = 1 \end{cases} \tag{4.28}$$

and

$$\begin{cases} r_{22} - t_2 r_{32} = \Omega_{12} \\ -r_{12} + t_1 r_{32} = \Omega_{22} \\ t_2 r_{12} - t_1 r_{22} = \Omega_{32} \\ r_{12}^2 + r_{22}^2 + r_{32}^2 = 1 \end{cases} \tag{4.29}$$

where r_{ij} and Ω_{ij} denote the ij-th elements of matrix $\boldsymbol{R}$ and $\boldsymbol{\Omega}$ respectively.

From (4.28) and (4.29), the first and second column vectors of matrix $\boldsymbol{R}$ can be calculated analytically, and the third column vector of $\boldsymbol{R}$ is then given by the cross product of these two columns. Then we obtain the solution for the rotation matrix.

II. An Indirect Solution By rearranging (4.16), we have

$$\boldsymbol{R}^{\mathrm{T}}\boldsymbol{C} - \boldsymbol{D} = \boldsymbol{0} \tag{4.30}$$

where $\boldsymbol{C} = [\boldsymbol{t}]_{\times}$ and $\boldsymbol{D} = \lambda \boldsymbol{H}^{\mathrm{T}}[\boldsymbol{t}]_{\times}$.

We find that the above equation has the same form as that of (2.16) derived by Weng et al. [170]. Therefore, similar steps can be taken to solve the rotation matrix.

Assuming $\boldsymbol{C} = [\boldsymbol{C}_1 \quad \boldsymbol{C}_2 \quad \boldsymbol{C}_3]$ and $\boldsymbol{D} = [\boldsymbol{D}_1 \quad \boldsymbol{D}_2 \quad \boldsymbol{D}_3]$, we define a 4×4 matrix as

$$\boldsymbol{B} = \sum_{i=1}^{3} \boldsymbol{B}_i^{\mathrm{T}} \boldsymbol{B}_i \tag{4.31}$$

where $\boldsymbol{B}_i = \begin{bmatrix} 0 & (\boldsymbol{C}_i - \boldsymbol{D}_i)^{\mathrm{T}} \\ \boldsymbol{D}_i - \boldsymbol{C}_i & [\boldsymbol{C}_i + \boldsymbol{D}_i]_{\times} \end{bmatrix}$.

Let $\boldsymbol{q}_1 = (q_0 \quad q_1 \quad q_2 \quad q_3)^{\mathrm{T}}$ be the eigenvector of $\boldsymbol{B}$ associated with the smallest singular value. The solution for $\boldsymbol{R}$ is given by

$$\boldsymbol{R} = \begin{bmatrix} q_0^2 + q_1^2 - q_2^2 - q_3^2 & 2(q_1 q_2 - q_0 q_3) & 2(q_1 q_3 + q_0 q_2) \\ 2(q_1 q_2 + q_0 q_3) & q_0^2 - q_1^2 + q_2^2 - q_3^2 & 2(q_2 q_3 - q_0 q_1) \\ 2(q_1 q_3 - q_0 q_2) & 2(q_2 q_3 + q_0 q_1) & q_0^2 - q_1^2 - q_2^2 + q_3^2 \end{bmatrix} \tag{4.32}$$

4.3.4 Implementation Procedure

Assuming the intrinsic parameters of the camera and the projector have been calibrated from static calibration stage, when the configuration of the system is changes, the procedure for the dynamic calibration and 3D reconstruction in our structured light system can be summarized as follows:

Step 1: Extracting four or more feature correspondences between the camera image and the light pattern of the projector;

Step 2: Computing the Homography matrix between the camera plane and projector plane according to (2.10);
Step 3: Determining the scale factor λ and its sign as proposed in Sect. 4.3.1;
Step 4: Calculating the translation vector using any one approach from Sect. 4.3.2;
Step 5: Calculating the rotation matrix using any one approach from Sect. 4.3.3;
Step 6: Optionally, the results can be refined by bundle adjustment, after having obtained the relative pose.
Step 7: Combining the coefficient matrix $\boldsymbol{X}$ and vector $\boldsymbol{x}$ in (4.3) and performing 3D reconstruction by (4.4).

Remark 1 Step 6 is optional. Since the corresponding features between the camera and the projector are abundant, we can establish a great lot of equations on the Homography matrix in Step 2. Furthermore, SVD method self is a least-square optimization method. Therefore usually without Step 6, satisfactory estimations can still be obtained and simultaneously the implementation is fast.

Remark 2 Once the rotation matrix and translation vector are calibrated, our method can be used for plane locating directly. It should be noted that the Homographic matrix $\overline{\boldsymbol{H}}$ and the scale factor λ will vary with different plane. Therefore, we are required to determine them firstly.

Assuming $\overline{\boldsymbol{H}}$ has been obtained, from (4.15), we have

$$\lambda\overline{\boldsymbol{H}} - \boldsymbol{R} = \boldsymbol{t}\boldsymbol{n}^{\mathrm{T}} \tag{4.33}$$

Since both vector $\boldsymbol{t}$ and vector $\boldsymbol{n}$ are rank one, the rank of the left side of (4.33) is also one, which means the determinant of any 2×2 sub-matrix will vanish. Therefore, 6 equations can be obtained for the scale factor λ, that is

$$\det{(\lambda\overline{\boldsymbol{H}} - \boldsymbol{R})_{2\times 2}} \tag{4.34}$$

The scale λ can be determined by any one of these equations or using all in the least squares sense.

Then, the normal vector of the plane can be directly given by

$$\boldsymbol{n}^{\mathrm{T}} = \boldsymbol{t}^{\mathrm{T}}(\lambda\overline{\boldsymbol{H}} - \boldsymbol{R}) \tag{4.35}$$

However, the normal vector of the plane can not be determined in this way if the translation vector is null, i.e. $\boldsymbol{t} = \boldsymbol{0}$. Fortunately, this case does not happen in our structured light system. If it does happen in a passive vision, we can firstly perform 3D reconstruction using (4.4) to reconstruct some points on the plane, and then compute its normal vector with these points.

Remark 3 In Step 7, the 3D reconstruction is performed via traditional triangulation method, which means the rays corresponding to the two matched pixels should intersect at a 3D point as shown in Fig. 4.1. In practice, various types of noise, such as distortions, feature extraction errors and small errors in the model parameters, will lead to those rays do not always intersect. So the problem is how to find a 3D point which optimally fits the measured pixels. Equation (4.4) provides a good solution in the least-square manner. Geometrically, it outputs a 3D point as the mid-point of the shortest transversal between the two rays. In the experiments, we find that satisfactory results can be obtained with the method.

4.4 Error Analyses

In this section, we present a typical sensitivity analysis on the calibration results. We assume that the translation vector and the rotation matrix are computed by (4.24) and (4.32) respectively. In the presence of noise, their covariance matrices with respect to those of the feature detection noises are derived step by step using the matrix perturbation theory.

4.4.1 *Errors in the Homographic Matrix*

In our structured light system, the projection point $\boldsymbol{m}_p$ is directly obtained from the predefined light pattern, which can be deemed as accurate. Various error sources, such as feature detection and localization errors, exist in the projection point $\boldsymbol{m}_c$ from the camera image. These errors will result in the errors in the estimated Homographic matrix. For simplicity, we assume that the projection points have been normalized before the estimations.

Assuming the i-th image point $\boldsymbol{m}_{c,i}$ has an error vector $\delta_{\boldsymbol{m}_{c,i}} = [\delta_{u_i} \quad \delta_{v_i} \quad 0]^{\mathrm{T}}$, from (2.10), we have the error matrix $\boldsymbol{\Delta}_A$ in the coefficient matrix $\boldsymbol{A}$:

$$\boldsymbol{\Delta}_A = \begin{bmatrix} \delta_{u_1} & \delta_{v_1} & 0 & 0 & 0 & 0 & -u'_1\delta_{u_1} & -u'_1\delta_{v_1} & 0 \\ 0 & 0 & 0 & \delta_{u_1} & \delta_{v_1} & 0 & -v'_1\delta_{u_1} & -v'_1\delta_{v_1} & 0 \\ \cdots & \cdots & \cdots & \cdots & \cdots & \cdots & \cdots & \cdots & \cdots \\ \delta_{u_n} & \delta_{v_n} & 0 & 0 & 0 & 0 & -u'_n\delta_{u_n} & -u'_n\delta_{v_n} & 0 \\ 0 & 0 & 0 & \delta_{u_n} & \delta_{v_n} & 0 & -v'_n\delta_{u_n} & -v'_n\delta_{v_n} & 0 \end{bmatrix}$$

Let $\boldsymbol{\delta}_{A^{\mathrm{T}}}$ represent the vector form of $\boldsymbol{\Delta}_{\mathrm{A}}^{\mathrm{T}}$ by the column first order. For simplicity, we assume that the errors between different image points and the two components in image coordinates are uncorrelated, and they have the same variance σ^2. With this assumption, the covariance matrix of $\boldsymbol{A}$ can be calculated by

$$\boldsymbol{\Gamma}_{A^{\mathrm{T}}} = E\left(\delta_A \delta_{\mathrm{A}}^{\mathrm{T}}\right) = \sigma^2 diag\left\{\boldsymbol{D}_1, \boldsymbol{D}_2, \cdots, \boldsymbol{D}_n\right\}$$

where

$$
\boldsymbol{D}_i = \begin{bmatrix} \boldsymbol{\Lambda} & \boldsymbol{0} & -u_i'\boldsymbol{\Lambda} & \boldsymbol{0} & \boldsymbol{\Lambda} & -v_i'\boldsymbol{\Lambda} \\ \boldsymbol{0} & \boldsymbol{0} & \boldsymbol{0} & \boldsymbol{0} & \boldsymbol{0} & \boldsymbol{0} \\ -u_i'\boldsymbol{\Lambda} & \boldsymbol{0} & u_i'^2\boldsymbol{\Lambda} & \boldsymbol{0} & -u_i'\boldsymbol{\Lambda} & u_i'v_i'\boldsymbol{\Lambda} \\ \boldsymbol{0} & \boldsymbol{0} & \boldsymbol{0} & \boldsymbol{0} & \boldsymbol{0} & \boldsymbol{0} \\ \boldsymbol{\Lambda} & \boldsymbol{0} & -u_i'\boldsymbol{\Lambda} & \boldsymbol{0} & \boldsymbol{\Lambda} & -v_i'\boldsymbol{\Lambda} \\ -v_i'\boldsymbol{\Lambda} & \boldsymbol{0} & u_i'v_i'\boldsymbol{\Lambda} & \boldsymbol{0} & -v_i'\boldsymbol{\Lambda} & v_i'^2\boldsymbol{\Lambda} \end{bmatrix}, \quad \boldsymbol{\Lambda} = \begin{bmatrix} 1 & 0 & 0 \\ 0 & 1 & 0 \\ 0 & 0 & 0 \end{bmatrix}
$$

and $diag\,\{\boldsymbol{D}_1, \cdots, \boldsymbol{D}_n\}$ denotes the diagonal matrix with the corresponding diagonal elements.

Since $\boldsymbol{Q}$ is defined as $\boldsymbol{Q} = \boldsymbol{A}^{\mathrm{T}}\boldsymbol{A}$ in (2.10). By the first order approximation, the error matrix $\boldsymbol{\Delta}_Q$ can be computed by

$$
\boldsymbol{\Delta}_Q = (\boldsymbol{A} + \boldsymbol{\Delta}_A)^{\mathrm{T}}(\boldsymbol{A} + \boldsymbol{\Delta}_A) - \boldsymbol{A}^{\mathrm{T}}\boldsymbol{A} \approx \boldsymbol{A}^{\mathrm{T}}\boldsymbol{\Delta}_A + \boldsymbol{A}\boldsymbol{\Delta}_A^{\mathrm{T}} \tag{4.36}
$$

Rewriting the error matrices in the column first order, we obtain

$$
\boldsymbol{\delta}_Q = (\boldsymbol{F}_Q + \boldsymbol{G}_Q)\boldsymbol{\delta}_{A^{\mathrm{T}}} \tag{4.37}
$$

where

$$
\boldsymbol{F}_Q = \begin{bmatrix} a_{11}\boldsymbol{I}_9 & a_{21}\boldsymbol{I}_9 & \cdots & a_{2n,1}\boldsymbol{I}_9 \\ a_{12}\boldsymbol{I}_9 & a_{22}\boldsymbol{I}_9 & \cdots & a_{2n,2}\boldsymbol{I}_9 \\ \cdots & \cdots & \cdots & \cdots \\ a_{19}\boldsymbol{I}_9 & a_{29}\boldsymbol{I}_9 & \cdots & a_{2n,9}\boldsymbol{I}_9 \end{bmatrix}, \quad \boldsymbol{G}_Q = \begin{bmatrix} a_{11} & \cdots & 0 & \vdots & a_{2n,1} & \cdots & 0 \\ \cdots & \cdots & \cdots & \vdots & \cdots & \cdots & \cdots \\ a_{19} & \cdots & 0 & \vdots & a_{2n,9} & \cdots & 0 \\ \vdots & \vdots & \vdots & \vdots & \vdots & \vdots & \vdots \\ 0 & \cdots & a_{11} & \vdots & 0 & \cdots & a_{2n,1} \\ \cdots & \cdots & \cdots & \vdots & \cdots & \cdots & \cdots \\ 0 & \cdots & a_{19} & \vdots & 0 & \cdots & a_{2n,9} \end{bmatrix},
$$

and a_{ij} represents the (i, j) element in matrix $\boldsymbol{A}$.

Let $\boldsymbol{v}_i$ and d_i be the i-th eigenvector and the corresponding eigenvalue of matrix $\boldsymbol{Q}$ respectively and $\boldsymbol{V} = [\boldsymbol{v}_1 \quad \boldsymbol{v}_2 \quad \cdots \quad \boldsymbol{v}_9]$. According to perturbation theory [170], the first order perturbation on the vector $\boldsymbol{h}$ is

$$
\begin{aligned}
\boldsymbol{\delta}_h &= \boldsymbol{V}\boldsymbol{\Lambda}_Q\boldsymbol{V}^{\mathrm{T}}\boldsymbol{\Delta}_Q\boldsymbol{v}_1 \\
&= \boldsymbol{V}\boldsymbol{\Lambda}_Q\boldsymbol{V}^{\mathrm{T}}[v_{11}\boldsymbol{I}_9 \quad v_{21}\boldsymbol{I}_9 \quad \cdots \quad v_{91}\boldsymbol{I}_9]\boldsymbol{\delta}_Q = \boldsymbol{D}_h\boldsymbol{\delta}_{A^{\mathrm{T}}}
\end{aligned} \tag{4.38}
$$

where:

$\boldsymbol{I}_9$ denotes the 9×9 identity matrix, v_{ij} means the i-th elements in $\boldsymbol{v}_j$,

$$\boldsymbol{\Lambda}_Q = diag\left\{0, (d_1 - d_2)^{-1}, \cdots, (d_1 - d_9)^{-1}\right\},$$

and $\boldsymbol{D}_h = \boldsymbol{V}\boldsymbol{\Lambda}_Q\boldsymbol{V}^{\mathrm{T}}\left[v_{11}\boldsymbol{I}_9 \quad v_{21}\boldsymbol{I}_9 \quad \cdots \quad v_{91}\boldsymbol{I}_9\right](\boldsymbol{F}_Q + \boldsymbol{G}_Q)$.

Equation (4.38) describes the error relationship between image point localization and the Homographic matrix. Once we obtain the error matrix $\boldsymbol{\Delta}_H$ on $\overline{\boldsymbol{H}}$, the error δ_{λ^2} on the eigenvalue λ^2 can be obtained as follows.

Let $\boldsymbol{S}_{\overline{H}} = \overline{\boldsymbol{H}\boldsymbol{H}}^{\mathrm{T}}$. To the first order approximation, the error matrix for $\boldsymbol{S}_{\overline{H}}$ can be computed as

$$\begin{aligned}\boldsymbol{\Delta}_{S_{\overline{H}}} &= \left(\overline{\boldsymbol{H}} + \boldsymbol{\Delta}_{\overline{H}}\right)\left(\overline{\boldsymbol{H}} + \boldsymbol{\Delta}_{\overline{H}}\right)^{\mathrm{T}} - \overline{\boldsymbol{H}\boldsymbol{H}}^{\mathrm{T}} \\ &\approx \overline{\boldsymbol{H}}\boldsymbol{\Delta}_{\overline{H}}^{\mathrm{T}} + \overline{\boldsymbol{H}}^{\mathrm{T}}\boldsymbol{\Delta}_{\overline{H}}\end{aligned} \tag{4.39}$$

Rewriting in the column first order, we have

$$\boldsymbol{\delta}_{S_{\overline{H}}} = \left(\boldsymbol{F}_{S_{\overline{H}}} + \boldsymbol{G}_{S_{\overline{H}}}\right)\boldsymbol{\delta}_h = \boldsymbol{D}_{S_{\overline{H}}}\boldsymbol{\delta}_h \tag{4.40}$$

where $$\boldsymbol{F}_{S_{\overline{H}}} = \begin{bmatrix} h_{11}\boldsymbol{I} & h_{21}\boldsymbol{I} & h_{31}\boldsymbol{I} \\ h_{12}\boldsymbol{I} & h_{22}\boldsymbol{I} & h_{32}\boldsymbol{I} \\ h_{13}\boldsymbol{I} & h_{23}\boldsymbol{I} & h_{33}\boldsymbol{I} \end{bmatrix},$$

$$\boldsymbol{G}_{S_{\overline{H}}} = \begin{bmatrix} h_{11} & 0 & 0 & h_{21} & 0 & 0 & h_{31} & 0 & 0 \\ h_{12} & 0 & 0 & h_{22} & 0 & 0 & h_{32} & 0 & 0 \\ h_{13} & 0 & 0 & h_{23} & 0 & 0 & h_{33} & 0 & 0 \\ 0 & h_{11} & 0 & 0 & h_{21} & 0 & 0 & h_{31} & 0 \\ 0 & h_{12} & 0 & 0 & h_{22} & 0 & 0 & h_{32} & 0 \\ 0 & h_{13} & 0 & 0 & h_{23} & 0 & 0 & h_{33} & 0 \\ 0 & 0 & h_{11} & 0 & 0 & h_{21} & 0 & 0 & h_{31} \\ 0 & 0 & h_{12} & 0 & 0 & h_{22} & 0 & 0 & h_{32} \\ 0 & 0 & h_{13} & 0 & 0 & h_{23} & 0 & 0 & h_{33} \end{bmatrix}$$

and h_{ij} represents the (i, j) element in matrix $\overline{\boldsymbol{H}}$.

Let $\boldsymbol{u}_2$ be the eigenvector corresponding to the second eigenvalue of matrix $\boldsymbol{S}_{\overline{H}}$. According to the perturbation theorem, the first order perturbation on λ^2 is

$$\boldsymbol{\delta}_{\lambda^2} = \boldsymbol{u}_2^{\mathrm{T}}\boldsymbol{\Delta}_{S_{\overline{H}}}\boldsymbol{u}_2 = \boldsymbol{u}_2^{\mathrm{T}}\left[u_{12}\boldsymbol{I} \quad u_{22}\boldsymbol{I} \quad u_{32}\boldsymbol{I}\right]\boldsymbol{\delta}_{S_{\overline{H}}}.$$

Then the relationship between δ_{λ^2} and $\delta_{A^{\mathrm{T}}}$ is

$$\boldsymbol{\delta}_{\lambda^2} = \boldsymbol{D}_{\lambda^2}\boldsymbol{\delta}_{A^{\mathrm{T}}} \tag{4.41}$$

where $\boldsymbol{D}_{\lambda^2} = u_2^{\mathrm{T}}\left[u_{12}\boldsymbol{I} \quad u_{22}\boldsymbol{I} \quad u_{32}\boldsymbol{I}\right]\boldsymbol{D}_{S_{\overline{H}}}\boldsymbol{D}_h$.

Assuming $\left|\frac{\delta_{\lambda^2}}{\lambda^2}\right| << 1$, the perturbation on λ is $\delta_\lambda = \sqrt{\lambda^2 + \delta_{\lambda^2}} - \lambda \approx \frac{\delta_{\lambda^2}}{2\lambda}$.

4.4.2 Errors in the Translation Vector

Let $\boldsymbol{I} = [\boldsymbol{e}_1 \quad \boldsymbol{e}_2 \quad \boldsymbol{e}_3]$. From the formula for matrix $\boldsymbol{W}$, its error matrix can be computed as

$$\boldsymbol{\Delta}_W = \boldsymbol{\Delta}_{S_{\overline{H}}} - \delta_{\lambda^2}\boldsymbol{I} \tag{4.42}$$

Rewriting in the column first order, we obtain

$$\boldsymbol{\delta}_W = \boldsymbol{\delta}_{S_{\overline{H}}} - \begin{bmatrix}\boldsymbol{e}_1^{\mathrm{T}} & \boldsymbol{e}_2^{\mathrm{T}} & \boldsymbol{e}_3^{\mathrm{T}}\end{bmatrix}^{\mathrm{T}} \boldsymbol{\delta}_{\lambda^2} = \boldsymbol{D}_W \boldsymbol{\delta}_{A^{\mathrm{T}}} \tag{4.43}$$

where $\boldsymbol{D}_W = \boldsymbol{D}_{S_{\overline{H}}} - \begin{bmatrix}\boldsymbol{e}_1^{\mathrm{T}} & \boldsymbol{e}_2^{\mathrm{T}} & \boldsymbol{e}_3^{\mathrm{T}}\end{bmatrix}^{\mathrm{T}} \boldsymbol{D}_{\lambda^2}$.

Once we obtain the error vector δ_W for the matrix $\boldsymbol{W}$, the errors δ_{t_1} and δ_{t_2} for the translation vector can be computed as follows.

Expanding t_1 in (4.24) using Taylor series for multiple variables, we have

$$\begin{aligned} t_1(w_{11}, w_{13}, w_{33}) \approx t_1 &- \frac{1}{2\sqrt{w_{13}^2 - w_{11}w_{33}}}\delta_{w_{11}} \\ &+ \left(\frac{1}{w_{33}} + \frac{w_{13}}{w_{33}\sqrt{w_{13}^2 - w_{11}w_{33}}}\right)\delta_{w_{13}} \\ &+ \left(\frac{w_{13}}{2w_{33}^2\sqrt{w_{13}^2 - w_{11}w_{33}}} - \frac{w_{13}}{w_{33}^2}\right)\delta_{w_{33}} \end{aligned}$$

Similarly, we can have the first order approximation for t_2 in (4.24). Then writing the first-order perturbations on t_1 and t_2 in matrix form, we have

$$\begin{bmatrix}\boldsymbol{\delta}_{t_1} \\ \boldsymbol{\delta}_{t_2}\end{bmatrix} = \begin{bmatrix}\boldsymbol{D}_{t_L} & \boldsymbol{D}_{t_R}\end{bmatrix} \boldsymbol{\delta}_W \tag{4.44}$$

where:

$$\boldsymbol{D}_{t_L} = \frac{1}{b_1^2}\begin{bmatrix} -\frac{\sqrt{-b_1^3}}{2} & 0 & \frac{b_1^2 + w_{13}\sqrt{-b_1^3}}{w_{33}} \\ w_{23}b_1 + w_{33}b_1t_2 - \frac{a_1\sqrt{-b_1}}{2} & (w_{33}t_1 - w_{13})b_1 & \frac{w_{13}a_1\sqrt{-b_1} - a_1b_1}{w_{33}} - 2w_{13}b_1t_2 - (w_{23}t_1 + w_{12})b_1 \end{bmatrix}$$

and

$$\boldsymbol{D}_{t_R} = \frac{1}{b_1^2}\begin{bmatrix} 0 & 0 & 0 & 0 & 0 & \frac{w_{11}\sqrt{-b_1^3} - 2w_{13}b_1^2}{2w_{33}^2} \\ 0 & 0 & (w_{11} - w_{13}t_1)b_1 & 0 & 0 & w_{12}t_1b_1 + w_{11}t_2b_1 + \frac{w_{11}a_1\sqrt{-b_1} + 2w_{13}a_1b_1}{2w_{33}^2} \end{bmatrix},$$

a_1 and b_1 are defined as in (4.23).

Then the error relationship between the translation vector and feature points can be described as

$$\begin{bmatrix} \boldsymbol{\delta}_{t_1} \\ \boldsymbol{\delta}_{t_2} \end{bmatrix} = \begin{bmatrix} \boldsymbol{D}_{t_L} & \boldsymbol{D}_{t_R} \end{bmatrix} \boldsymbol{D}_W \boldsymbol{\delta}_{A^{\mathrm{T}}} = \boldsymbol{D}_t \boldsymbol{\delta}_{A^{\mathrm{T}}} \tag{4.45}$$

where $\boldsymbol{D}_t = \begin{bmatrix} \boldsymbol{D}_{t_L} & \boldsymbol{D}_{t_R} \end{bmatrix} \boldsymbol{D}_W$.

4.4.3 Errors in the Rotation Matrix

For conciseness, we define a new vector $\boldsymbol{\rho}$ that combines λ, t_1 and t_2 with the vector $\boldsymbol{h}$, such that $\boldsymbol{\rho} = [\lambda \quad t_1 \quad t_2 \quad \boldsymbol{h}^{\mathrm{T}}]^{\mathrm{T}}$.

Considering (4.38), (4.41) and (4.45), we have the error vector for $\boldsymbol{\rho}$ as

$$\boldsymbol{\delta}_{\rho} = \boldsymbol{D}_{\rho} \boldsymbol{\delta}_{A^{\mathrm{T}}} \tag{4.46}$$

where $\boldsymbol{D}_{\rho} = \begin{bmatrix} \frac{\boldsymbol{D}_{\lambda^2}}{(2\lambda)} \\ \boldsymbol{D}_t \\ \boldsymbol{D}_h \end{bmatrix}$.

To the first order approximation, the error matrix for the matrix $\boldsymbol{B}$ in (4.31) is

$$\begin{aligned} \boldsymbol{\Delta}_B &= \sum_{i=1}^{3} \left(\boldsymbol{B}_i + \boldsymbol{\Delta}_{B_i}\right)^{\mathrm{T}} \left(\boldsymbol{B}_i + \boldsymbol{\Delta}_{B_i}\right) - \sum_{i=1}^{3} \boldsymbol{B}_i^{\mathrm{T}} \boldsymbol{B}_i \\ &\approx \sum_{i=1}^{3} \left(\boldsymbol{B}_i^{\mathrm{T}} \boldsymbol{\Delta}_{B_i} + \boldsymbol{\Delta}_{B_i^{\mathrm{T}}} \boldsymbol{B}_i\right) \end{aligned} \tag{4.47}$$

Rewriting in the column first order, the relationship between δ_B and δ_ρ is

$$\boldsymbol{\delta}_B = 2\boldsymbol{D}_B \boldsymbol{\delta}_\rho \tag{4.48}$$

where $\boldsymbol{D}_B$ is derived in appendix B.

Let $\boldsymbol{q}_i$ and γ_i be the i-th eigenvector and corresponding eigenvalue of the matrix $\boldsymbol{B}$ and $\boldsymbol{\eta} = \begin{bmatrix} \boldsymbol{q}_1 & \boldsymbol{q}_2 & \boldsymbol{q}_3 & \boldsymbol{q}_4 \end{bmatrix}$. Using the matrix perturbation theorem, the first order perturbation on $\boldsymbol{q}_1$ is

$$\boldsymbol{\delta}_{q_1} = \boldsymbol{\eta} \boldsymbol{\Lambda}_B \boldsymbol{\eta}^{\mathrm{T}} \boldsymbol{\Delta}_B \boldsymbol{q}_1 = \boldsymbol{D}_{q_1} \boldsymbol{\delta}_\rho \tag{4.49}$$

where q_i is the i-th element of q_1,$\boldsymbol{D}_{q_1} = 2\boldsymbol{\eta} \boldsymbol{\Lambda}_B \boldsymbol{\eta}^{\mathrm{T}} \begin{bmatrix} q_0 \boldsymbol{I}_4 & q_1 \boldsymbol{I}_4 & q_2 \boldsymbol{I}_4 & q_3 \boldsymbol{I}_4 \end{bmatrix} \boldsymbol{D}_B$ and

$$\boldsymbol{\Lambda}_B = diag\left\{0, (\gamma_1 - \gamma_2)^{-1}, (\gamma_1 - \gamma_3)^{-1}, (\gamma_1 - \gamma_4)^{-1}\right\}.$$

Let the Jacobian matrix of the rotation matrix be

$$J_q = 2\begin{bmatrix} q_0 & q_3 & -q_2 & -q_3 & q_0 & q_1 & q_2 & -q_1 & q_0 \\ q_1 & q_2 & q_3 & q_2 & -q_1 & q_0 & q_3 & -q_0 & -q_1 \\ -q_2 & q_1 & -q_0 & q_1 & q_2 & q_3 & q_0 & q_3 & -q_2 \\ -q_3 & q_0 & q_1 & -q_0 & -q_3 & q_2 & q_1 & q_2 & q_3 \end{bmatrix}^{\mathrm{T}}.$$

In the column first order, the error on the rotation matrix can be computed as

$$\boldsymbol{\delta}_R = \boldsymbol{J}_q \boldsymbol{\delta}_q = \boldsymbol{D}_R \boldsymbol{\delta}_{A^{\mathrm{T}}} \tag{4.50}$$

where $\boldsymbol{D}_R = \boldsymbol{J}_q \boldsymbol{D}_{q_1} \boldsymbol{D}_\rho$.

In summary, the error vectors of the translation vector and rotation matrix are expressed in terms of linear transformation of the errors of the coefficient matrix $\boldsymbol{A}$ in (2.10). Once they are obtained, their covariance matrices can be calculated as

$$\boldsymbol{\Gamma}_t = E(\boldsymbol{\delta}_t \boldsymbol{\delta}_t^{\mathrm{T}}) = \boldsymbol{D}_t \boldsymbol{\Gamma}_{A^{\mathrm{T}}} \boldsymbol{D}_t^{\mathrm{T}} \tag{4.51}$$

and

$$\boldsymbol{\Gamma}_R = E(\boldsymbol{\delta}_R \boldsymbol{\delta}_R^{\mathrm{T}}) = \boldsymbol{D}_R \boldsymbol{\Gamma}_{A^{\mathrm{T}}} \boldsymbol{D}_R^{\mathrm{T}} \tag{4.52}$$

where $E(\cdot)$ denotes the expectation. The uncertainty for the localization of image points can be obtained via an experimental procedure although its analytical expression is difficult to determine. In practice, the procedure can be performed with two steps: (1) image acquisition of the scene; (2) localization of the feature points. The performance is repeated several times and the covariance matrix can be evaluated with the standard deviations of the pixel coordinates.

Up to now, we have analyzed the error relationships between the pose parameters and the image point in terms of covariance matrices. With these relationships, the accuracy of the calibration results can be determined in advance given the uncertainty of the image point localization. Another potential use of the relationship is to provide a guideline for setting the system's configuration which can be expected to produce reasonable results. For example, high accuracy of feature detection and localization is required if we intend to estimate the relative pose with a good dispersion.

4.5 Experiments Study

4.5.1 Computer Simulation

4.5.1.1 Ambiguity of the Solutions

As to the ambiguity of the solutions, Tsai and Huang [120] and Longuet-Higgins [123] showed how the two possible interpretations of the camera motion could be determined in a closed form from the correspondences in two images of a planar surface. Negahdaripour [160, 161] determined the relationship between the two solutions in a closed form. In addition, the derivation demonstrated the explicit relationship between the ambiguity associated with planar scenes and that associated with curved surfaces. Knowledge of the explicit relationship between the two interpretations permits the calculation of one solution directly from the other. In this experiment, we will show that the results from our method coincide with their conclusions by simulations.

In the simulated system, we assumed that the intrinsic parameters of both the camera and the projector had been calibrated in static calibration stage. For each repeated experiment, the translation vector and three rotation angles of the rotation matrix were set randomly and the normal vector of the planar surface was also selected arbitrarily in order to cover all cases in practice. Virtual grid light pattern and camera image were generated for analysis. The resolution of the image was limited to be 600×400, and that of the pattern was 120×120. By randomly selecting n ($n > 4$) points on the object plane, their projection points in the camera image and the projector image were used to compute the Homographic matrix. Then the solutions for the translation vector and the rotation matrix were obtained from the proposed procedure. Here, 10000 random simulations were done to reveal the ambiguity of the solutions. It should be noted that multiple solutions are obtained by simply solving the given equations then discarding the complex ones. In order to find which choice corresponded to the true configuration, the cheirality constraint [162] (The constraint that the scene points should be in front of the camera and the projector.) was imposed. In the first test, we used the algebraic solution in Sect. 4.3.2 and the direct solution in Sect. 4.3.3 to solve the translation vector and the rotation matrix. The final results were shown in Table 4.1 for the distribution of solutions after applying the cheirality constraint in the simulations. Figure 4.2 illustrated the graphs of the distributions from Table 4.1. From these data, we can see that there are one or two solutions in most cases, and the probability is about 95.65%. However, as many as four solutions were obtained in some trials. Then in another test, we used the formulas (4.24) and (4.27) in Sect. 5.3.2 and (4.32) in Sect. 5.3.3 to solve the translation vector and the rotation matrix, respectively. Unique solution was obtained in all the experiments. Here, a twisted pair was treated as one solution. However, the convention as to whether a twisted pair counts as one solution or two solutions varies from authors to authors (See page 21 in [163]). Therefore, these experiments demonstrate that the results from our method coincide with their conclusions. Furthermore, these experiments reveal the advantage of using (4.24) or (4.27) for the translation vector and (4.32) for the rotation matrix. So in the following experiments, we employed (4.24) and (4.32) to solve the translation vector and the rotation matrix since explicit expressions are given for the solutions.

4.5.1.2 Robustness of the Dynamic Calibration

In this example, the translation vector and three rotation angles between the camera's and the projector's coordinate systems were set to be $\boldsymbol{t}_0 = [3, 8, 1]$ and $\boldsymbol{r}_0 = [-\text{pi}/8\ \text{pi}/$

Table 4.1 Distributions of number of solutions

Num. of Solutions	Frequency
1	5305
2	4260
3	143
4	279

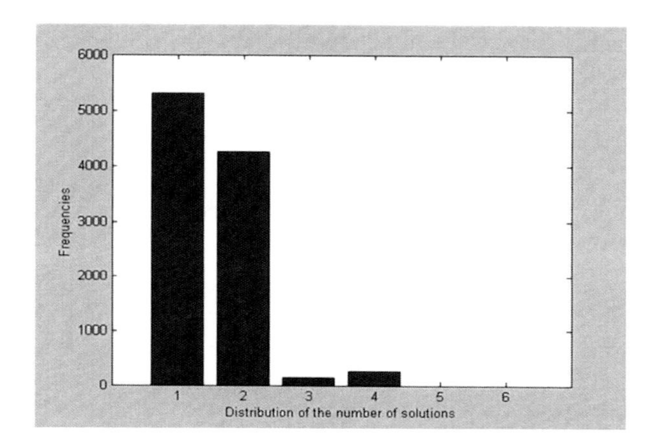

Fig. 4.2 Distributions of number of solutions from Table 4.1

5 pi/4]. The object was assumed to be a plane whose equation is $Z = 0.1X + 0.6Y + 560$ (Fig. 4.3). By randomly selecting 20 illuminated points on the scene plane, the correspondences on the camera plane and projector plane were identified first and used for solving the image-to-image Homography and then the relative pose parameters with the proposed method.

To test the robustness of the procedure, the residual errors, which is the discrepancies between theoretical values and computed results, are evaluated. Here, different levels of Gaussian noise, i.e. $N(0, \sigma^2)$, were added to the projection points. We varied the noise level from 0.0 to 1.0 pixel. For each noise level, we performed 50 trials and calculated the average residual errors of the rotation angles and translation vector. Since a twisted pair of solutions is obtained, the minimum residual error for the rotation angles and translation vectors is defined as follows

$$\min_i \left(\left\| \frac{\boldsymbol{r}_i}{\|\boldsymbol{r}_i\|} - \frac{\boldsymbol{r}_0}{\|\boldsymbol{r}_0\|} \right\| \right) \quad \text{and} \quad \min_i \left(\left\| \frac{\boldsymbol{t}_i}{\|\boldsymbol{t}_i\|} - \frac{\boldsymbol{t}_0}{\|\boldsymbol{t}_0\|} \right\| \right) \tag{4.53}$$

where $\boldsymbol{r}_i$ and $\boldsymbol{t}_i$ are the calculated solutions.

For comparison, we implemented Higgins's method [123] under the same condition since this method also gave analytic solutions for the pose problem by analyzing the plane-based Homography from two perspective views. Here, the mean value and standard deviation of minimum residual errors from experimental results were computed. The results of the rotation angles and translation vector were shown in

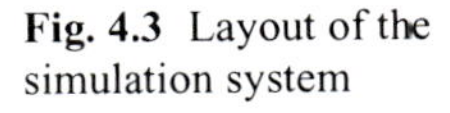

Fig. 4.3 Layout of the simulation system

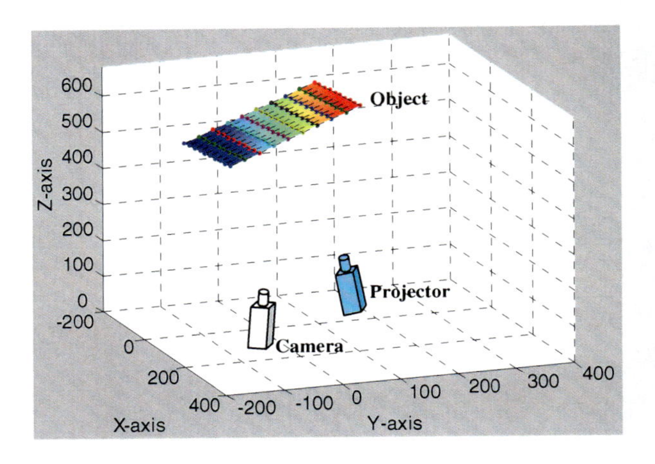

Figs. 4.4 and 4.5, where data1 and data2 represented the results from our method and Higgins's method respectively. From these figures, we can see that both methods work well when the noise is less than 0.5 pixel. When the noise level increases, the residual errors increase too and the graphs become somewhat bumpy. This can be improved if we either increase the number of points or choose those points more carefully (e.g. when the points are more evenly distributed). Furthermore, to reduce the bumpy effect, bundle adjustment can be used to improve the results obtained from the analytic solution. On the whole, our method outperforms Higgins's method in the presence of noise data. For example, when the noise level was 0.5 pixel, the average error residuals of our method were $1.41 \pm 0.86\%$ and $2.41 \pm 1.06\%$ for the rotation angles and translation vector, while the results from Higgins's method were $2.19 \pm 3.11\%$ and $3.46 \pm 2.27\%$. In our method, an over-constrained system can be easily constructed to increase the robustness. Hence, the standard deviation for translation vector in our method is smaller, as seen in Fig. 4.4b. We notice that the estimates of the translation vector are a bit more sensitive to noise than the estimates of rotation. This observation has already been discussed by other authors, e.g. [164]. However, our method shows its robustness for the translation vector since an over-constrained system is employed.

In the simulations, we also compared our method with the classical eight-point algorithm [106] where the fundamental matrix was computed and then decomposed directly into rotation matrix and translation vector. We observed that the residual errors from our method and the eight-point algorithm were in a similar range, which validated our method. In this work, our motivation is to recalibrate the vision system using planar information, in which case the eight-point algorithm will fail as it requires those points in general 3D positions. Furthermore, the minimum number of points required in the eight-point algorithm is eight while four points is sufficient in our method.

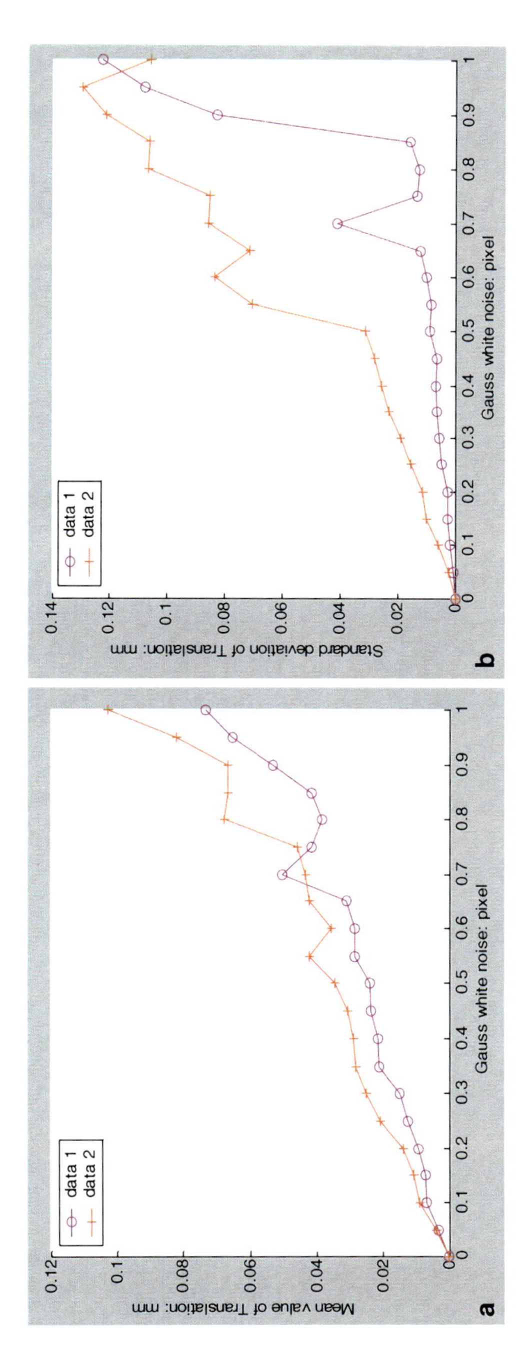

Fig. 4.4 Translation vectors vs. noise levels. **a** Average results of the residual errors. **b** Standard deviation of the residual errors

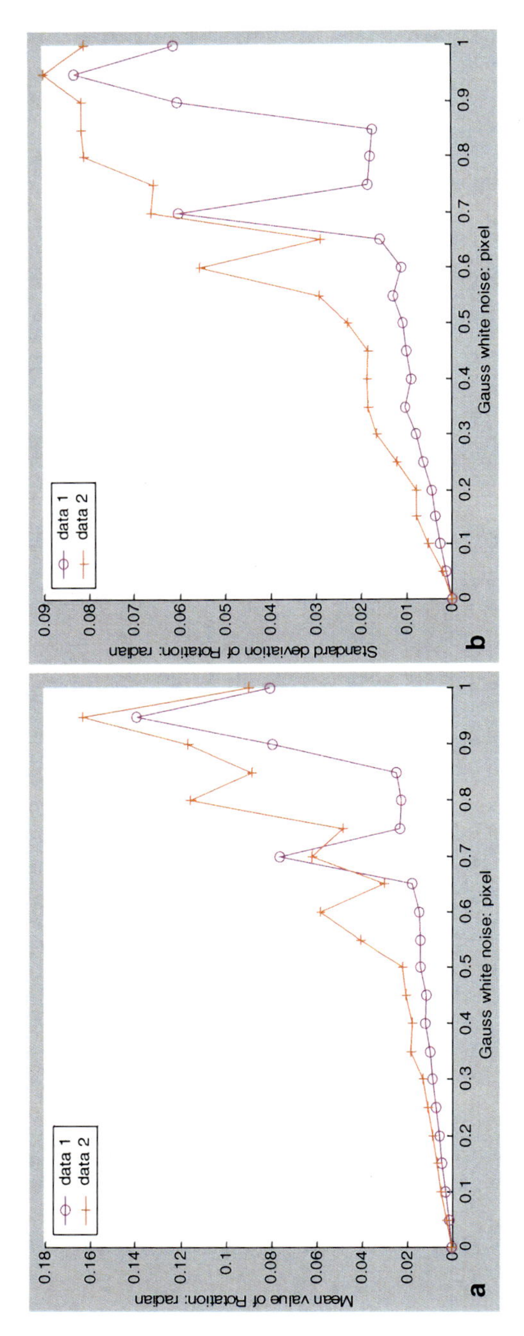

Fig. 4.5 Rotation angles vs. noise levels. **a** Mean values of the residual errors. **b** Standard deviation of the residual errors

4.5.1.3 Validity of the Sensitivity Analysis

In this experiment, we verify the validity and accuracy of our method on the sensitivity analysis, where the covariance matrices for the translation vector and rotation matrix are computed according to (4.51) and (4.52). For clarity, we define the predicted errors as the squared roots of the traces of those matrices, i.e. $\sqrt{trace(\Gamma_t)}$ and $\sqrt{trace(\Gamma_R)}$. Here, Gaussian noise ranged from 0.0 to 1.0 pixel was added to the projection points. For each noise level, 50 trials were implemented to obtain the predicted errors and the residual errors for the translation vector and the rotation matrix. In general, the closer they are, the higher accuracy the prediction technique would be. The experimental results for the translation vector and the rotation matrix were illustrated in Figs. 4.6a and b respectively, where the average discrepancies between the residual errors and the predicted errors were plotted. From these figures, we can see that their discrepancies become larger quickly when the noise level is increased. This is mainly due to the first order approximation involved in the method. However, when the level of noise is low, say below 0.5 pixel, very small discrepancies are observed. Therefore, the predicted errors are very close to the real values which indicate that the sensitivity analysis is valid and accurate.

In the last simulation, we carried out some experiments on the residual errors and the predicted errors for different pose parameters. Since the rotation parameters do not significantly affect the solutions, we just test on different translation vectors. Here, the directions of the translation vector were changed from [3, 0, 1] to [0, 8, 1] with twenty evenly distributed directions. Random Gaussian noise with 0.5 pixel variance was added to the feature points. For each direction, ten trials were performed and the average results for the rotation matrix and translation vector were shown in Figs. 4.7a and b, where data1 and data2 represented the residual errors and the predicted errors respectively. As can be seen from these figures, the predicted errors are strongly correlated with the residual errors which indicate their discrepancies are very small. The first several trials were more sensitive to noise, hence more unreliable, e.g. the errors for both the rotation matrix and the translation vector in the second, the fourth and the fifth trials were larger than the others.

4.5.2 Real Data Experiment

The structured light system used in this experiment mainly consists of two components, i.e. a PULNIX TMC-9700 CCD color camera and a PLUS V131 DLP projector (Fig. 4.8). The projector, which is controlled by the light pattern as designed in Sect. 2.2.2 of Chap. 2, projects a bundle of encoded color light into the scene. Each light stripe in the pattern, intersecting with the scene, will produce a deformed curve according to different surface shape. The camera captures the illuminated scene in a RGB image, which is used for dynamic calibration and 3D reconstruction. Here, the projector is considered as a virtual camera, which actively

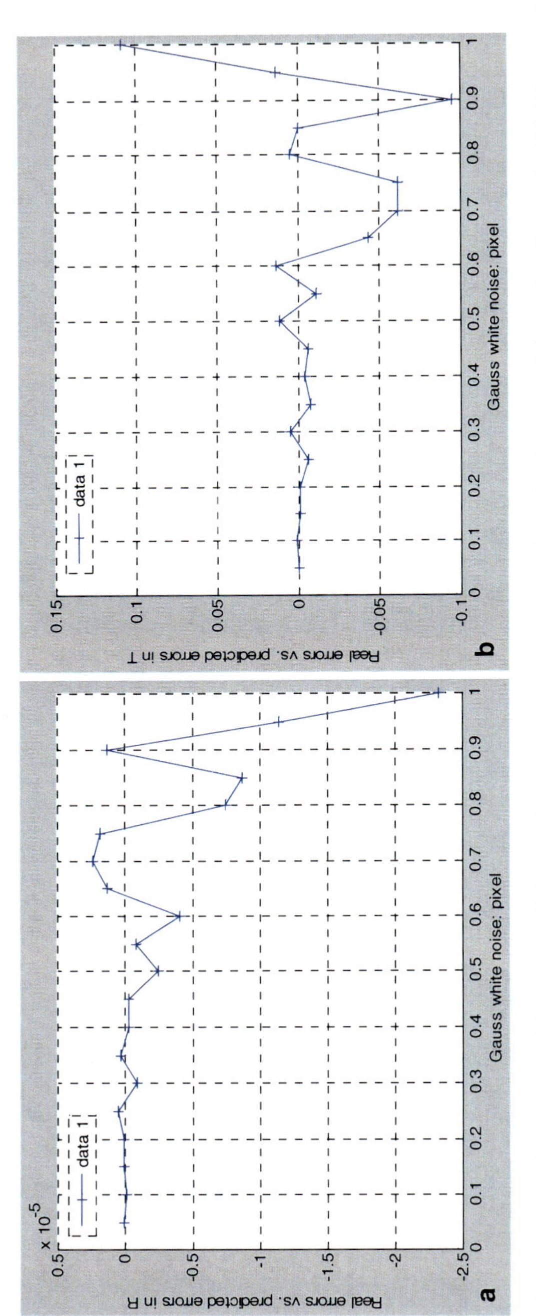

Fig. 4.6 Test on the accuracy of the error prediction method. **a** Discrepancies between the residual errors and the predicted errors in $\boldsymbol{R}$. **b** Discrepancies between the residual errors and the predicted errors in $\boldsymbol{t}$

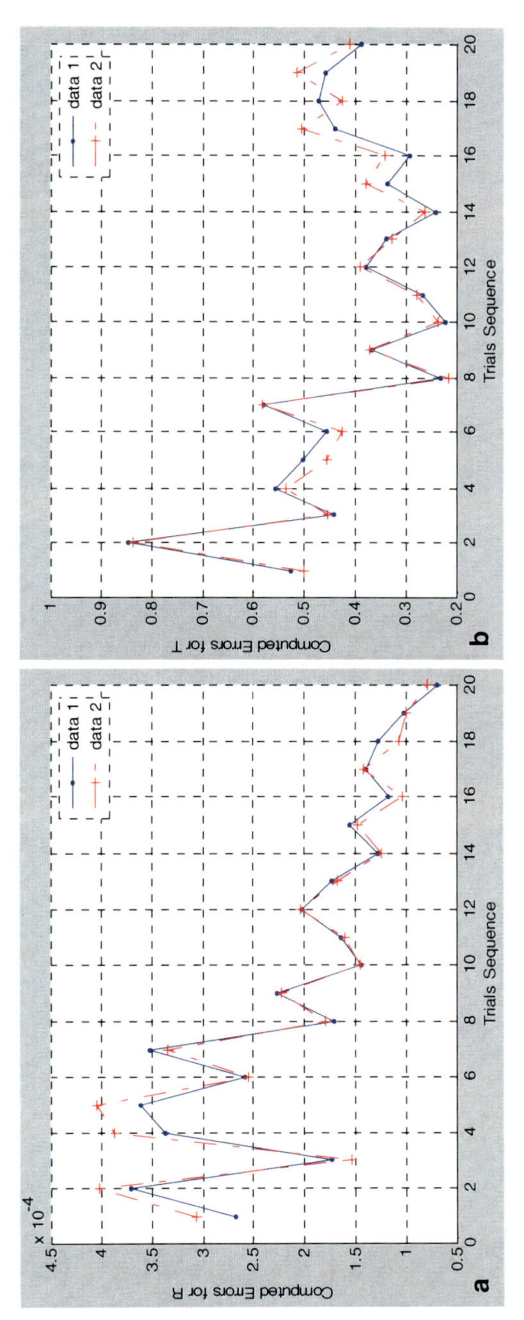

Fig. 4.7 Test on different translation parameters. **a** Experimental results for $\boldsymbol{R}$. **b** Experimental results for $\boldsymbol{t}$

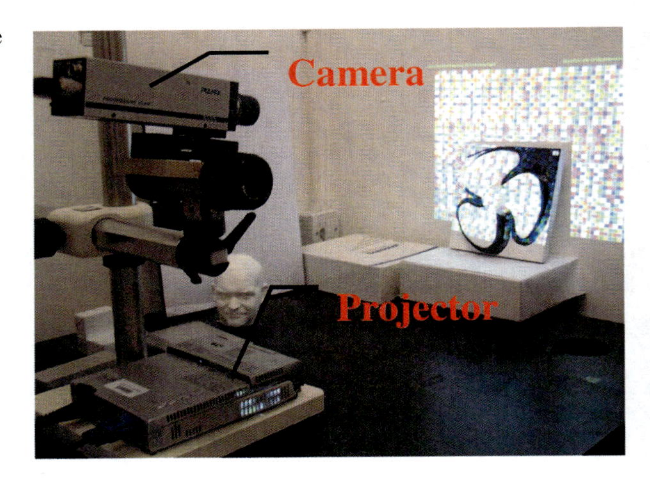

Fig. 4.8 Configuration of the structured light system

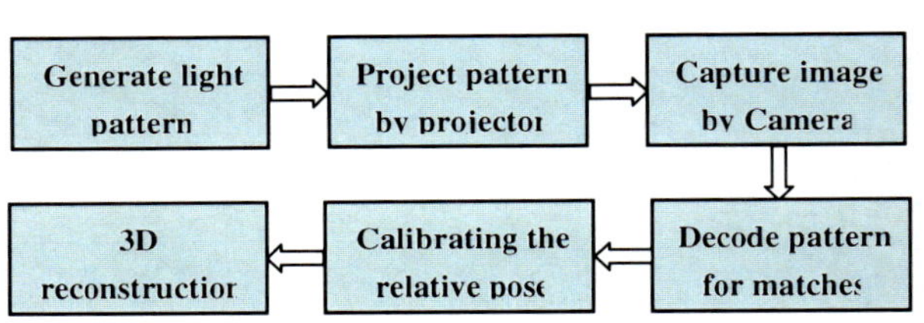

Fig. 4.9 Flow Chart of the Structured light System

emits rather than passively accepts those light rays. Hence, the system is used as an active stereoscopic vision system.

The implementation flow chart is shown in Fig. 4.9. During the experiment, an arbitrary area of the wall in our lab was deemed as the planar structure. We assumed that the intrinsic parameters of the camera were calibrated using the proposed method in Chap. 3. As to the projector, we first obtain a set of 3D points (acting as Control Points) and their projection points in the projector plane. Then the direct linear transformation method (DLT) from Abdel-Aziz and Karara [87] was implemented to compute the projection matrix. To extract the intrinsic parameters of the projector from the matrix, a technique from [165] was adopted. For simplicity, the skew factors of the camera and the projector were set to be zeroes and the results for the intrinsic parameters were given in Table 4.2. When calibrating the extrinsic parameters, four or more point correspondences from the wall were chosen between the projector plane and the camera image. The proposed procedure was performed with these correspondences. Here were some calibration data where the Homographic vector was $\boldsymbol{h}=[-0.4047,\ 1.0547,\ -0.3501,\ 1.2877,\ 0.0416,\ 0.0373,\ 0.2386,\ -0.1385,\ 1.0356]$, and then the rotation angles and translation vector were given in Table 4.3.

After the vision system had been calibrated, we performed 3-D object reconstruction to test the calibration results. Figure 4.10a gave an image of a man's head model. Totally, 226 points from the model were reconstructed. Here, the polygonized results of the reconstructed point clouds were shown from two different viewpoints as

Table 4.2 Intrinsic parameters obtained from static calibration

Intrinsic Para.	f_u	f_v	u_0	v_0
Camera	1215.3	1180.6	299.1	224.2
Projector	94.8811	89.1594	44.3938	25.4487

Table 4.3 Relative Pose in the vision system

Rotation Angles (radian)	Translation Vector (mm)
[− 2.1691, 2.1397, 0.3903]	[50.1919, − 28.8095, 1]

in Figs. 4.10e and f. Since this system is not configured precisely and no attempt has been made to obtain the ground truth, we do not know real values of the variable parameters of the system and the 3D point clouds. To visualize their accuracy, the reconstructed points were back-projected to the projector plane and the camera image using the calibrated results. For comparison, the original feature points and back-projected points were illustrated as in Figs. 4.10b and c. Figure 4.10d showed a zoomed part of the image for graceful view. Furthermore, the discrepancies between the back-projected points and the original features were estimated where the mean values for the projector points and image points were obtained as (0.2064, 0.3748) and (0.0272, 0.0163) respectively. From these figures and the quantitative results, we can see that the original feature points and the back-projected points are very close to each other as expected. This demonstrates that the relative pose calibrated by our method is valid and accurate.

In yet another experiment, we tested on a duck model as shown in Fig. 4.11a. Totally 143 points from the model were reconstructed. The polygonized results of the reconstructed point clouds were also shown from two different viewpoints as in Figs. 4.11e and f. To evaluate their accuracy, the point clouds were back-projected to the image plane and the projector plane respectively as in the last test. Figures 4.11b and c illustrated the experimental results. Figure 4.11d gave one zoomed part of the image. Quantitatively, the mean results of the discrepancies were (0.1829, 0.3336) and (0.0244, 0.0142). Again, it can be see that the original feature points and the re-projected points are very close to each other as expectably, which further demonstrates the validity and accuracy of our method.

4.6 Summary

This chapter presents the work on the relative pose problem with the assumption of one arbitrary plane in the structured light system. The image-to-image Homographic matrix is extensively explored to provide an analytic solution for the relative pose between the camera and the projector. Redundancy in the data can be easily incorporated to improve the reliability of the estimations in the presence of noise. The proposed system will recalibrate itself automatically when moving from one place to another or the configuration is changed, which allows an immediate 3D reconstruction to be followed.

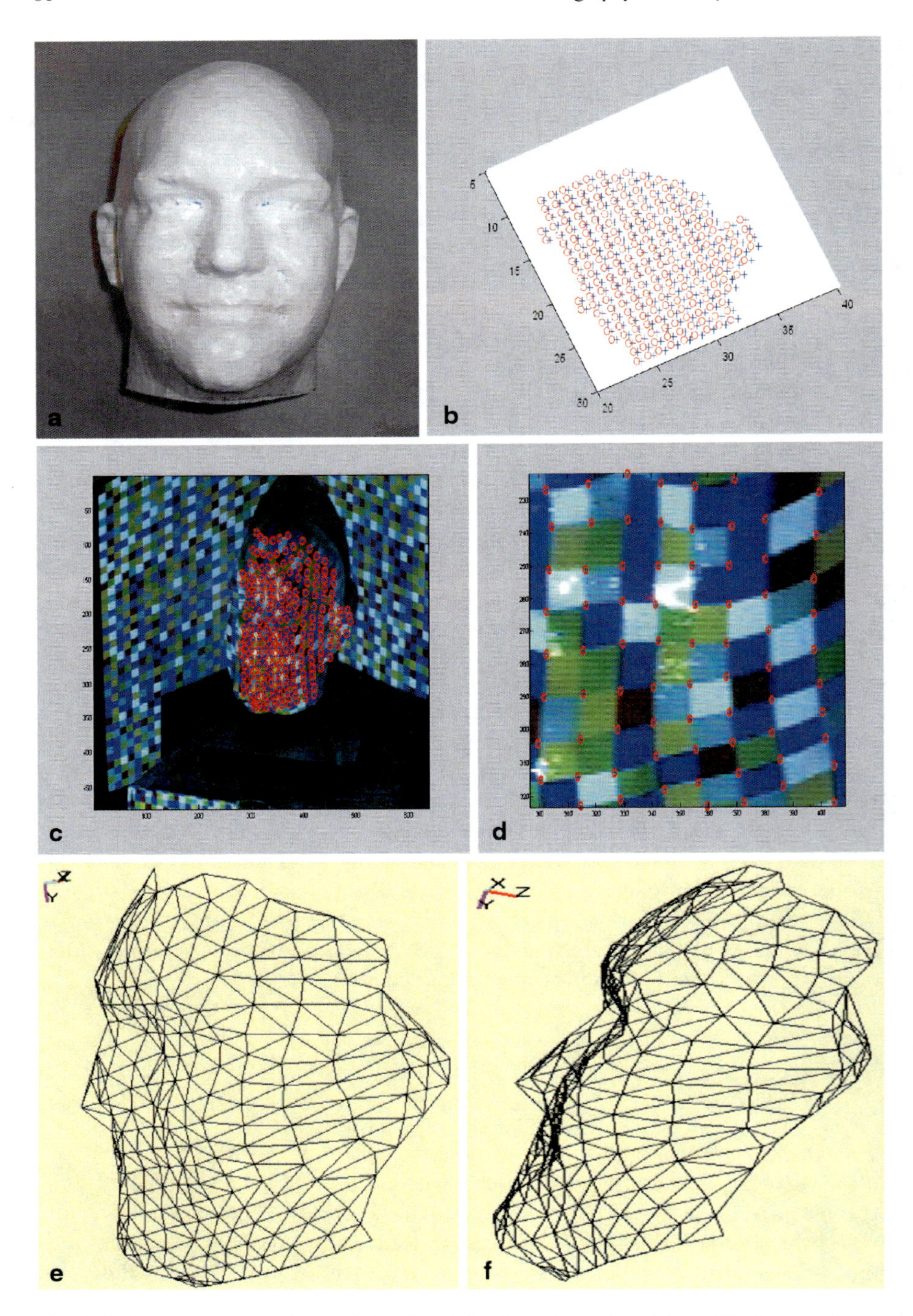

Fig. 4.10 An experiment on the man's head model. **a** A man's head model used for the experiment. **b**, **c**, **d** Original feature points and back-projected points: *Blue* '+' represents original feature points while *red* '*o*' represents back-projected points from reconstructed 3D points. They should coincide with each other theoretically. **e**, **f** Polygonized results of the points clouds from two viewpoints

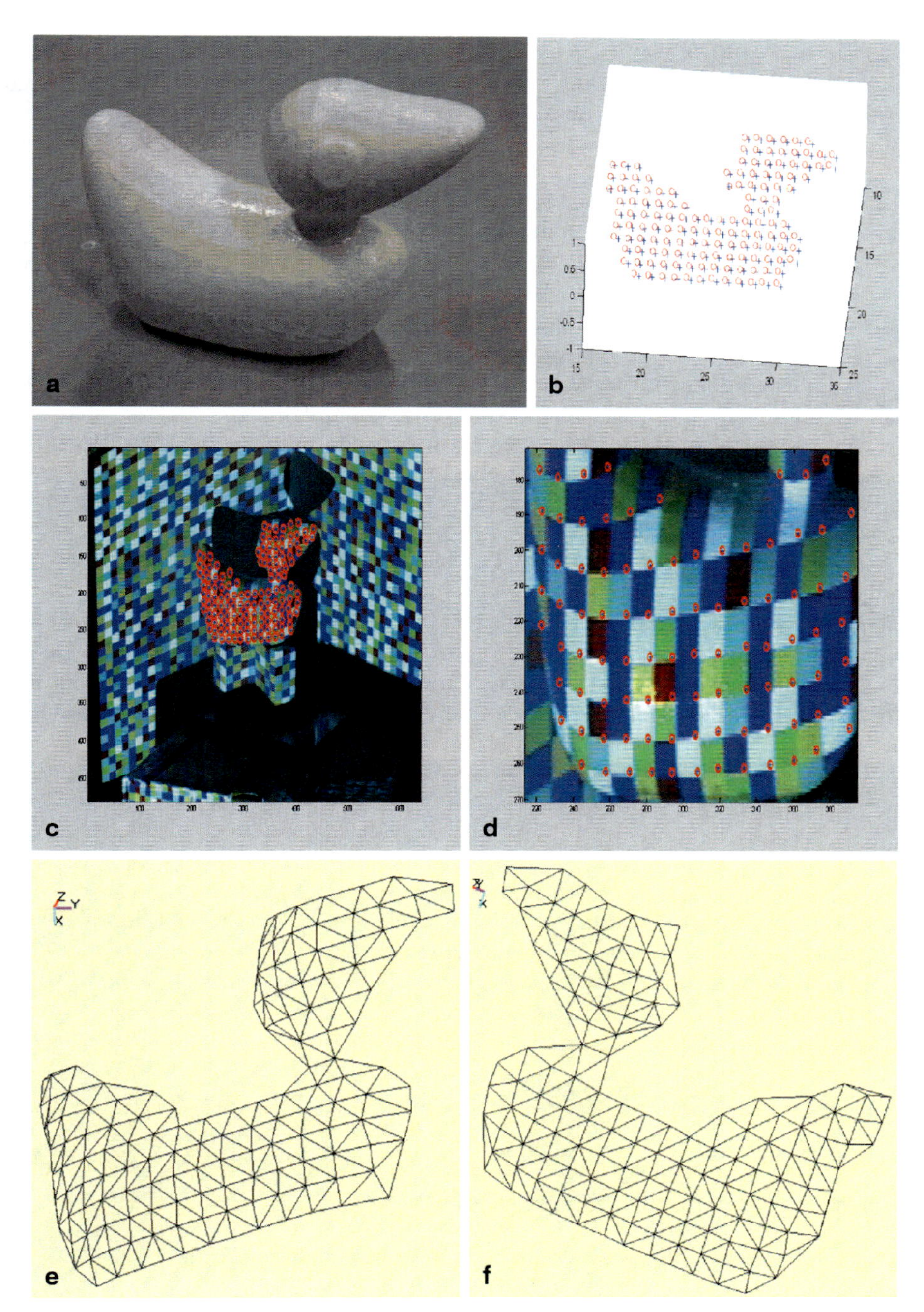

Fig. 4.11 Another experiment on a duck model. **a** A duck model used for the experiment. **b**, **c**, **d** Original feature points and back-projected points: *Blue* '+' represents original feature points while *red* '*o*' represents back-projected points from reconstructed 3D points. They should coincide with each other theoretically. **e**, **f** Polygonized results of the points clouds from two viewpoints

Based on the matrix perturbation theory, sensitivity analysis and error propagation from the noise in the image points to the estimated pose parameters are derived and validated. These results provide a means for studying the reliability and robustness of the solutions and a guideline for setting the system's configuration which can be expected to produce reasonable results in practical applications.

This method can be used in a static as well as a dynamic environment since only a single image is needed. When the scene is textureless, the projector will produce many feature points on the scene's surface with the pattern projection which are used for the dynamic calibration. However, this method is not restricted to structured light systems. It can be adopted for passive stereo vision if a textured environment is present. The experimental results show that this method is conceptually simple and can be efficiently implemented with acceptable results. Besides, the assumption of an arbitrary plane are easily satisfied in many practical applications since the plane can be a real or virtual one as long as it provides four or more coplanar points that enable the estimation of the Homographic matrix.

In practice, the planar or near planar scenes are encountered frequently in robotic tasks, e.g. road surface or ground plane in mobile robot navigation, walls or ceilings of a building for a climbing robot. The traditional methods, e.g. the eight-point algorithm [103], will fail or give a poor performance since they require a pair of images from the general three-dimensional scene. The existing techniques using planar information, e.g. Zhang and Hanson [124], mainly depend on noise-free images and solve for the exact solutions which will be fragile in the presence of noise. In these cases, our method shows its advantages and provides a good solution since an over-constrained system is constructed from the Homographic matrix directly.

Appendix A

In this appendix, we show that three linear equations can be obtained from the six quadratic equations by polynomial eliminations.

From w_{23}*(4.22.1)-w_{33}*(4.22.5), we get

$$w_{33}w_{13}t_1t_2 + (w_{33}w_{12} - 2w_{23}w_{13})t_1 - w_{33}w_{11}t_2 + w_{23}w_{11} = 0 \tag{A.1}$$

From w_{22}*(4.22.1)-w_{33}*(4.22.6), we get

$$2w_{33}w_{12}t_1t_2 - w_{33}w_{11}t_2^2 - 2w_{22}w_{13}t_1 + w_{22}w_{11} = 0 \tag{A.2}$$

From w_{13}*(4.22.2)-w_{33}*(4.22.4), we have

$$w_{23}w_{33}t_1t_2 - w_{22}w_{33}t_1 + (w_{12}w_{33} - 2w_{23}w_{13})t_2 + w_{22}w_{13} = 0 \tag{A.3}$$

From (A.2) + w_{11}*(4.22.2), we have

$$w_{12}w_{33}t_1t_2 - w_{22}w_{13}t_1 - w_{11}w_{23}t_2 + w_{11}w_{22} = 0 \tag{A.4}$$

From w_{13}*(4.22.3)-(A.1), we have

$$(w_{13}w_{23} - w_{12}w_{33})t_1 + \left(w_{11}w_{33} - w_{13}^2\right)t_2 + w_{12}w_{13} - w_{11}w_{23} = 0 \qquad \text{(A.5)}$$

From w_{23}*(4.22.3)-(A.3), we have

$$\left(w_{22}w_{33} - w_{23}^2\right)t_1 + (w_{13}w_{23} - w_{12}w_{33})t_2 + w_{12}w_{23} - w_{13}w_{22} = 0 \qquad \text{(A.6)}$$

From w_{12}*(4.22.3)-(A.4), we have

$$(w_{22}w_{13} - w_{12}w_{23})t_1 + (w_{11}w_{23} - w_{12}w_{13})t_2 + w_{12}^2 - w_{11}w_{22} = 0 \qquad \text{(A.7)}$$

We then have obtained the three linear equations in (A.5), (A.6) and (A.7).

Appendix B

Let c_{ij} and d_{ij} be the (i, j) elements of matrix $\boldsymbol{C}$ and $\boldsymbol{D}$ respectively.

Let $\boldsymbol{D_B} = [\boldsymbol{D}_\lambda \quad \boldsymbol{D}_{t_1} \quad \boldsymbol{D}_{t_2} \quad \boldsymbol{D}_{h_L} \quad \boldsymbol{D}_{h_R}]$. Then its five elements can be formulated as follows.

$$\boldsymbol{D}_\lambda = \begin{bmatrix} d_{31}t_2 - d_{32}t_1 + 2c_{13}t_2 + d_{23}t_1 - 2c_{12} - d_{21} - 2c_{23}t_1 + d_{12} - d_{13}t_2 \\ d_{22}t_1 - d_{31} - d_{21}t_2 + d_{33}t_1 \\ -d_{12}t_1 + d_{11}t_2 + d_{33}t_2 - d_{32} \\ -d_{13}t_1 - d_{23}t_2 + d_{11} + d_{22} \\ -d_{31}t_2 + d_{32}t_1 + 2c_{13}t_2 - d_{23}t_1 - 2c_{12} + d_{21} - 2c_{23}t_1 + d_{12} - d_{13}t_2 \\ d_{13}t_1 - d_{23}t_2 - d_{11} + d_{22} \\ -d_{12}t_1 + d_{11}t_2 - d_{33}t_2 + d_{32} \\ -d_{31}t_2 + d_{32}t_1 + 2c_{13}t_2 + d_{23}t_1 - 2c_{12} - d_{21} - 2c_{23}t_1 - d_{12} + d_{13}t_2 \\ -d_{22}t_1 - d_{31} + d_{21}t_2 + d_{33}t_1 \\ d_{31}t_2 - d_{32}t_1 + 2c_{13}t_2 - d_{23}t_1 - 2c_{12} + d_{21} - 2c_{23}t_1 - d_{12} + d_{13}t_2 \end{bmatrix},$$

$$D_{t_1} = \begin{bmatrix} c_{13}h_4 + c_{23}h_5 - d_{13}h_4 - 2c_{23}\lambda - d_{23}h_5 - d_{33}h_6 - c_{12}h_7 + c_{23}h_9 \\ \quad +d_{22}h_8 + d_{12}h_7 + d_{32}h_9 + d_{23}\lambda - d_{32}\lambda \\ -c_{23}h_8 + d_{22}\lambda + c_{23}h_6 + d_{33}\lambda \\ -c_{13}h_6 - d_{12}\lambda + c_{12}h_9 + c_{23}h_7 \\ -c_{23}h_4 + c_{13}h_5 - d_{13}\lambda - c_{12}h_8 \\ c_{13}h_4 - c_{23}h_5 - d_{13}h_4 - 2c_{23}\lambda - d_{23}h_5 - d_{33}h_6 - c_{12}h_7 - c_{23}h_9 \\ \quad +d_{22}h_8 + d_{12}h_7 + d_{32}h_9 - d_{23}\lambda + d_{32}\lambda \\ c_{23}h_4 + c_{13}h_5 + d_{13}\lambda - c_{12}h_8 \\ c_{13}h_6 - d_{12}\lambda - c_{12}h_9 + c_{23}h_7 \\ -c_{13}h_4 + c_{23}h_5 - d_{13}h_4 - 2c_{23}\lambda - d_{23}h_5 - d_{33}h_6 + c_{12}h_7 - c_{23}h_9 \\ \quad +d_{22}h_8 + d_{12}h_7 + d_{32}h_9 + d_{23}\lambda \\ +d_{32}\lambda c_{23}h_8 - d_{22}\lambda + c_{23}h_6 + d_{33}\lambda \\ -c_{13}h_4 - c_{23}h_5 - d_{13}h_4 - 2c_{23}\lambda - d_{23}h_5 - d_{33}h_6 + c_{12}h_7 + c_{23}h_9 \\ \quad +d_{22}h_8 + d_{12}h_7 + d_{32}h_9 - d_{23}\lambda - d_{32}\lambda \end{bmatrix},$$

$$D_{t_2} = \begin{bmatrix} -c_{13}h_1 + d_{33}h_3 + d_{23}h_2 + 2c_{13}\lambda - d_{11}h_7 - c_{13}h_9 - d_{21}h_8 + d_{13}h_1 \\ \quad -c_{12}h_8 - c_{23}h_2 - d_{31}h_9 - d_{13}\lambda + d_{31}\lambda \\ c_{13}h_8 - c_{23}h_3 - c_{12}h_9 - d_{21}\lambda \\ d_{33}\lambda + c_{13}h_3 + c_{11}\lambda - c_{13}h_7 \\ -c_{13}h_2 + c_{12}h_7 - d_{23}\lambda + c_{23}h_1 \\ -c_{13}h_1 + d_{33}h_3 + d_{23}h_2 + 2c_{13}\lambda - d_{11}h_7 + c_{13}h_9 - d_{21}h_8 + d_{13}h_1 \\ \quad +c_{12}h_8 + c_{23}h_2 - d_{31}h_9 - d_{13}\lambda - d_{31}\lambda \\ -c_{13}h_2 - c_{12}h_7 - d_{23}\lambda - c_{23}h_1 \\ -d_{33}\lambda - c_{13}h_3 + c_{11}\lambda - c_{13}h_7 \\ c_{13}h_1 + d_{33}h_3 + d_{23}h_2 + 2c_{13}\lambda - d_{11}h_7 + c_{13}h_9 - d_{21}h_8 + d_{13}h_1 \\ \quad -c_{12}h_8 - c_{23}h_2 - d_{31}h_9 + d_{13}\lambda - d_{31}\lambda \\ -c_{13}h_8 - c_{23}h_3 - c_{12}h_9 + d_{21}\lambda \\ c_{13}h_1 + d_{33}h_3 + d_{23}h_2 + 2c_{13}\lambda - d_{11}h_7 - c_{13}h_9 - d_{21}h_8 + d_{13}h_1 \\ \quad +c_{12}h_8 + c_{23}h_2 - d_{31}h_9 + d_{13}\lambda + d_{31}\lambda \end{bmatrix},$$

$$
\boldsymbol{D}_{h_L} = \begin{bmatrix}
(d_{13} - c_{13})t_2 - d_{12} + c_{12} & (d_{23} - c_{23})t_2 - d_{22} & d_{33}t_2 - d_{32} - c_{23} & (c_{13} - d_{13})t_1 + d_{11} & (-d_{23} + c_{23})t_1 + d_{21} + c_{12} \\
0 & c_{23} & -c_{23}t_2 & 0 & -c_{13} \\
-c_{23} & 0 & c_{13}t_2 - c_{12} & c_{13} & 0 \\
c_{23}t_2 & c_{12} - c_{13}t_2 & 0 & -c_{23}t_1 - c_{12} & c_{13}t_1 \\
(d_{13} - c_{13})t_2 - d_{12} + c_{12} & (d_{23} + c_{23})t_2 - d_{22} & d_{33}t_2 - d_{32} + c_{23} & (c_{13} - d_{13})t_1 + d_{11} & (-d_{23} - c_{23})t_1 + d_{21} - c_{12} \\
-c_{23}t_2 & c_{12} - c_{13}t_2 & 0 & c_{23}t_1 + c_{12} & c_{13}t_1 \\
-c_{23} & 0 & -c_{13}t_2 + c_{12} & c_{13} & 0 \\
(d_{13} + c_{13})t_2 - d_{12} - c_{12} & (d_{23} - c_{23})t_2 - d_{22} & d_{33}t_2 - d_{32} + c_{23} & (-c_{13} - d_{13})t_1 + d_{11} & (-d_{23} + c_{23})t_1 + d_{21} + c_{12} \\
0 & -c_{23} & -c_{23}t_2 & 0 & c_{13} \\
(d_{13} + c_{13})t_2 - d_{12} - c_{12} & (d_{23} + c_{23})t_2 - d_{22} & d_{33}t_2 - d_{32} - c_{23} & (-c_{13} - d_{13})t_1 + d_{11} & (-d_{23} - c_{23})t_1 + d_{21} - c_{12}
\end{bmatrix},
$$

$$
\boldsymbol{D}_{h_R} = \begin{bmatrix}
-d_{33}t_1 + d_{31} + c_{13} & d_{12}t_1 - d_{11}t_2 - c_{12}t_1 & d_{22}t_1 - d_{21}t_2 - c_{12}t_2 & (d_{32} + c_{23})t_1 - (d_{31} + c_{13})t_2 \\
c_{23}t_1 + c_{12} & 0 & -c_{23}t_1 + c_{13}t_2 & -c_{12}t_2 \\
-c_{13}t_1 & c_{23}t_1 - c_{13}t_2 & 0 & c_{12}t_1 \\
0 & c_{12}t_2 & -c_{12}t_1 & 0 \\
-d_{33}t_1 + d_{31} - c_{13} & d_{12}t_1 - d_{11}t_2 - c_{12}t_1 & d_{22}t_1 - d_{21}t_2 + c_{12}t_2 & (d_{32} - c_{23})t_1 - (d_{31} - c_{13})t_2 \\
0 & -c_{12}t_2 & -c_{12}t_1 & 0 \\
c_{13}t_1 & c_{23}t_1 - c_{13}t_2 & 0 & -c_{12}t_1 \\
-d_{33}t_1 + d_{31} - c_{13} & d_{12}t_1 - d_{11}t_2 + c_{12}t_1 & d_{22}t_1 - d_{21}t_2 - c_{12}t_2 & (d_{32} - c_{23})t_1 - (d_{31} - c_{13})t_2 \\
c_{23}t_1 + c_{12} & 0 & c_{23}t_1 - c_{13}t_2 & -c_{12}t_2 \\
-d_{33}t_1 + d_{31} + c_{13} & d_{12}t_1 - d_{11}t_2 + c_{12}t_1 & d_{22}t_1 - d_{21}t_2 + c_{12}t_2 & (d_{32} + c_{23})t_1 - (d_{31} + c_{13})t_2
\end{bmatrix}.
$$

Chapter 5
3D Reconstruction with Image-to-World Transformation

In the previous chapter, the DLP projector is used as a virtual camera and the structured light system is treated as a stereoscopic vision system. In this chapter, we take the projector as a collection of light planes and study the possible algorithms for 3D Euclidean reconstruction with such a system. The plane-based formulation is explored to take advantage of the light planes in the pattern projection. The computational complexity of the algorithm is also analyzed and compared with the traditional triangulation method. The image-to-world transformation is employed for the 3D Euclidean reconstruction considering two different cases, i.e. two-known-plane case and one-known-plane case. Finally, some examples and experimental results are presented using the suggested system.

5.1 Introduction

In the structured light system, the image-to-world transformation is defined as a linear mapping of back projecting an image pixel to its corresponding world point. In the literature, many researchers have studied approaches for determining it. For example, Fig. 5.1 shows two scenarios for estimating this transformation. The left figure shows an intensity image of a corner in a room where there exists an arrangement of orthogonal planes whose equations expressed in the world coordinates have been premeasured accurately in Reid's work [68]. A plane-to-point method is suggested there for solving the eight unknown parameters. The right one is used by Huynh in [69] with a point-to-point algorithm, containing at least twelve known points with special distributions. When calibrating, those points should be visible to the system.

The objective of this work is to extend the above scenarios to a bit more general cases. In this chapter, we will discuss two different cases. One is to assume that there exist two planes in the scene and the other is with just one scene plane. Figure 5.2 presents one of the simulated configurations. Here, the projector is controlled by a grid light pattern, where each edge together with the projector center expands as a plane in 3D space with the name of light plane. Hence, it can be modeled as a collection of light planes. The object to be reconstructed is placed in front of the

B. Zhang, Y. F. Li, *Automatic Calibration and Reconstruction for Active Vision Systems*, Intelligent Systems, Control and Automation: Science and Engineering, DOI 10.1007/978-94-007-2654-3_5,

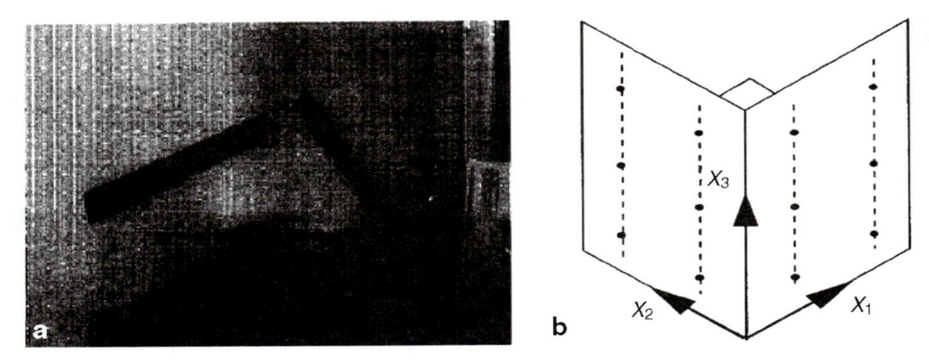

Fig. 5.1 **a** The intensity image of the calibration scene. **b** The calibration target and the coordinate system used for calibration

Fig. 5.2 A diagram of the active vision system

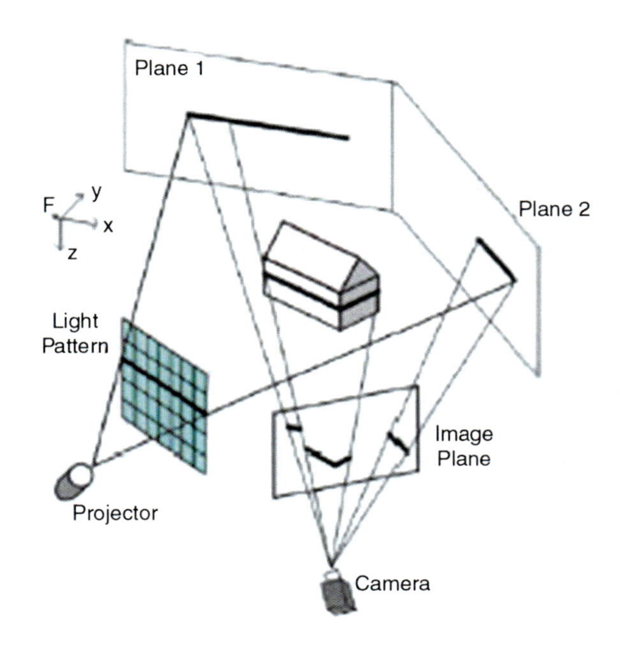

scene planes. Each light plane produces a deformed curve on the scene's surface according to the geometric shape and the camera captures these curves in an image. Since the scene planes can be larger enough, the problem of visibility in such a system is trivial.

When the equations of these planes are pre-calibrated at an offline stage, we show that the 3D structure of the object can be recovered without knowing the intrinsic or extrinsic parameters of the camera in this case. Furthermore, the system can recalibrate itself automatically whenever these parameters are changed. Herein, the

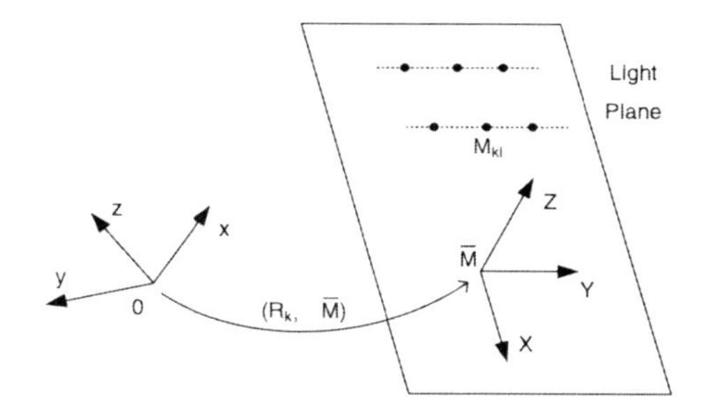

Fig. 5.3 Relationship between the world frame and the light plane frame

image-to-world transformation is investigated to take advantage of the light planes in the pattern projection. Next, we will introduce the mathematical model, taking one of the horizontal light planes as an example.

5.2 Image-to-World Transformation Matrix

We first define a world coordinate frame F_w with (x, y, z) representing its three coordinate axes in the scene. Let $\overline{\boldsymbol{M}} = (\overline{M_1} \quad \overline{M_2} \quad \overline{M_3})^{\mathrm{T}}$ be a given point on the k-th light plane $\boldsymbol{\Pi}_k$. We can define a local frame $F_{3\Pi k}$ originated at $\overline{\boldsymbol{M}}$, whose XY-axes lie on the plane and Z-axis aligns with its normal vector (Fig. 5.3). Next, we will discuss the image-to-world transformation matrix and the reconstruction principle in the active vision system.

The normal vector of $\boldsymbol{\Pi}_k$ is assumed to be $\boldsymbol{n}_k = (n_{k1} \quad n_{k2} \quad n_{k3})^{\mathrm{T}}$ in the world frame F_w. The direction of the z-axis of the frame F_w is $\boldsymbol{z} = (0 \quad 0 \quad 1)^{\mathrm{T}}$. Then the angle θ_k between $\boldsymbol{n}_k$ and the z-axis is simply the arc cosine of the inner product of the two vectors:

$$\theta_k = \cos^{-1}(\boldsymbol{n}_k{}^{\mathrm{T}}\boldsymbol{z}) = \cos^{-1}(n_{k3}) \tag{5.1}$$

and the unit vector perpendicular to the two vectors is

$$\boldsymbol{r}_k = \frac{(\boldsymbol{n}_k \times \boldsymbol{z})}{\|\boldsymbol{n}_k \times \boldsymbol{z}\|} \tag{5.2}$$

The rotation matrix from the world frame F_w to the light plane frame $F_{3\Pi k}$ can be given by Rodrigues formula (Pavlidis [158], pp. 366–368):

$$\boldsymbol{R}_k = \cos\theta_k \cdot \boldsymbol{I} + (1 - \cos\theta_k)\boldsymbol{r}_k\boldsymbol{r}_k^{\mathrm{T}} + \sin\theta_k \cdot [\boldsymbol{r}_k]_\times \tag{5.3}$$

If an arbitrary point on the light plane is denoted by $\boldsymbol{M}_{ki} = (M_{k1} \quad M_{k2} \quad M_{k3})^{\mathrm{T}}$ in the world frame F_w, its coordinates in light plane frame $F_{3\Pi k}$ will be

$$\boldsymbol{P}_{ki} = \boldsymbol{R}_k\boldsymbol{M}_{ki} - \boldsymbol{R}_k\overline{\boldsymbol{M}} \tag{5.4}$$

Let $\widetilde{\boldsymbol{M}}_{ki} = (M_{k1} \quad M_{k2} \quad M_{k3} \quad 1)^{\mathrm{T}}$ be the homogeneous form of the point. From (5.4), we will have

$$\widetilde{\boldsymbol{P}}_{ki} = \boldsymbol{T}_k \widetilde{\boldsymbol{M}}_{ki} \tag{5.5}$$

where $\boldsymbol{T}_k = \begin{bmatrix} \boldsymbol{R}_k & -\boldsymbol{R}_k \overline{\boldsymbol{M}} \\ \boldsymbol{0}^T & 1 \end{bmatrix}$ is a 4×4 matrix. It transforms a point from the world frame F_w to the light plane frame $F_{3\Pi k}$ (Fig. 5.3). It should be noted that $\widetilde{\boldsymbol{P}}_{ki}$ has the following coordinate form

$$\widetilde{\boldsymbol{P}}_{ki} = (m_{k1} \quad m_{k2} \quad 0 \quad 1)^{\mathrm{T}} \tag{5.6}$$

Let $F_{2\Pi k}$ be a 2D coordinate frame of the light plane and also originate at $\overline{\boldsymbol{M}}$. The X'-axis and Y'-axis of $F_{2\Pi k}$ coincide with those of $F_{3\Pi k}$. Then the coordinate relation between the world coordinate frame F_w and that of the light plane $F_{2\Pi k}$ follows

$$\boldsymbol{S}^{\mathrm{T}} \widetilde{\boldsymbol{m}}_{ki} = \boldsymbol{T}_k \widetilde{\boldsymbol{M}}_{ki} \tag{5.7}$$

where $\widetilde{\boldsymbol{m}}_{ki} = (m_{k1} \quad m_{k2} \quad 1)^{\mathrm{T}}$ is the corresponding homogenous representation of $\widetilde{\boldsymbol{P}}_{ki}$ expressed in the 2D frame $F_{2\Pi k}$ and $\boldsymbol{S} = \begin{bmatrix} 1 & 0 & 0 & 0 \\ 0 & 1 & 0 & 0 \\ 0 & 0 & 0 & 1 \end{bmatrix}$.

We can rewrite (5.7) as follows

$$\widetilde{\boldsymbol{m}}_{ki} = \boldsymbol{T}_{2k} \widetilde{\boldsymbol{M}}_{ki} \tag{5.8}$$

where $\boldsymbol{T}_{2k} = \boldsymbol{S}\boldsymbol{T}_k$ is the transformation from the world frame F_w to that of the light plane $F_{2\Pi k}$.

Let the Homographic matrix between the image plane and the light plane be denoted by $\boldsymbol{H}_k$. Let $\widetilde{\boldsymbol{m}}_i$ be the corresponding projection point of $\widetilde{\boldsymbol{m}}_{ki}$ in the camera image. Then we have

$$\widetilde{\boldsymbol{m}}_{ki} = \lambda_i \boldsymbol{H}_k \widetilde{\boldsymbol{m}}_i \tag{5.9}$$

Combining (5.9) and (5.7), we obtain

$$\boldsymbol{T}_k \widetilde{\boldsymbol{M}}_{ki} = \lambda_i \boldsymbol{S}^{\mathrm{T}} \boldsymbol{H}_k \widetilde{\boldsymbol{m}}_i \tag{5.10}$$

That is

$$\widetilde{\boldsymbol{M}}_{ki} = \lambda_i \boldsymbol{T}_{wk} \widetilde{\boldsymbol{m}}_i \tag{5.11}$$

where

$$\boldsymbol{T}_{wk} = \boldsymbol{T}_k^{-1} \boldsymbol{S}^{\mathrm{T}} \boldsymbol{H}_k \tag{5.12}$$

Here, $\boldsymbol{T}_{wk}$ is called the image-to-world transformation. It transforms a pixel in the image frame to its corresponding 3D point on the k-th light plane relative to the world coordinate system F_w. From (5.11), once this transformation matrix is recovered, 3D reconstruction is straightforward, i.e. by simply matrix multiplication. It is a 4×3 matrix and can be determined up to a nonzero scale. Therefore, it has eleven degrees of freedom.

5.3 Two-Known-Plane Based Method

Based on the above discussion, we can see that the image-to-world transformation is an important means for 3D data acquisition. To our knowledge, only a few works have been done on it. They are from Chen and Kak [67], Reid [68] and Huynh et al. [69] respectively employing a line-to-point, plane-to-point and point-to-point method. These techniques are limited to static calibration since they require either special devices or calibration targets with precisely known Euclidean structure. Any adjustment of their systems, e.g. auto-focus or movement of the camera, will result in a repeating calibration process and not be allowed during performing the vision tasks. However, this is frequently necessary in a dynamic environment with moving objects. Therefore, a more flexible vision system is desirable in practice.

From Eq. (5.12), we can see that $\boldsymbol{T}_{wk}$. has three different components, i.e. the transformation $\boldsymbol{T}_k$, constant matrix $\boldsymbol{S}$ and the Homography $\boldsymbol{H}_k$. In our system configuration, we assume that the light planes of the projector are calibrated and kept fixed. Therefore, the transformation $\boldsymbol{T}_k$ will keep constant once it is recovered.

On the other hand, adjustment on both intrinsic and extrinsic parameters of the camera, e.g. adaptive focusing or motion of the camera, should be allowed to achieve a better performance in different tasks. As a result, the Homographic matrix $\boldsymbol{H}_k$ will vary with different cases and need to be calibrated on-line.

In summary, the calibration tasks involve two stages. The first stage obtains the equations of the scene planes and the light planes. This is a static calibration and need to be performed only once. The second stage determines the online Homographic matrix and then the image-to-word transformation. This is a dynamic calibration and required to be performed whenever the system is changed.

5.3.1 *Static Calibration*

I. Determining the Scene Planes To determine the relationships between the two scene planes and the camera, a planar pattern with known structure is placed on each of the planes respectively. We define a local coordinate system attached for each of the pattern in the scene planes. Figure 5.4 illustrates a sketch for the scenario. Assuming that the world coordinate system coincides with the first local system, the equation for the first scene plane is $z = 0$. With the calibrated camera, we give a brief procedure for determining the equation of the second scene plane from its image:

1. Selecting some feature points on the first planar pattern and extracting their corresponding pixels in the image;
2. Estimating the homographic transformation matrix using those correspondences;
3. Computing the rotation matrix and translation vector between the first local system (the world system) and the camera system, say $\boldsymbol{R}_1$ and $\boldsymbol{t}_1$;
4. Repeating the above three steps to get $\boldsymbol{R}_2$ and $\boldsymbol{t}_2$ between the second local system and the camera system;

Fig. 5.4 An illustration for determining the scene planes

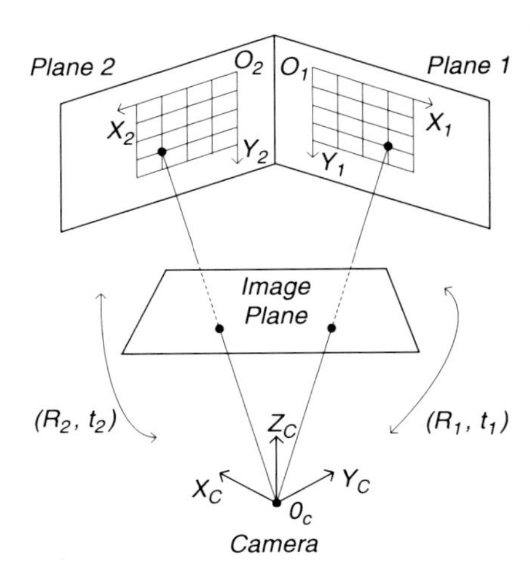

5. Let the vector $(\boldsymbol{n}^{\mathrm{T}} \quad d)^{\mathrm{T}}$ represent the equation of the second plane in the camera system. Then $\boldsymbol{n}$ is the third column of $\boldsymbol{R}_2$ and $d = \boldsymbol{n}^{\mathrm{T}} \cdot \boldsymbol{t}_2$.
6. Transforming the vector by $\begin{bmatrix} \boldsymbol{R}_1 & \boldsymbol{t}_1 \\ \boldsymbol{0}^{\mathrm{T}} & 1 \end{bmatrix}^{-\mathrm{T}}$, then we get the equation for the second plane expressed in the world system.

II. Determining the Light Planes Obviously, the equations of the light planes cannot be determined in the same way as the scene plane. As a plane can be determined by three points in general positions, this task can be done through finding these points. To improve the accuracy, we use a planar pattern containing many evenly distributed circles as a calibration target to produce a number of known points in the world coordinate system, instead of four known noncoplanar sets of three collinear world points used by Huynh et al. [69]. The centers of these circles can be easily identified and form sets of lines labelled by $\boldsymbol{L}_i$ (Fig. 5.5a). The planar pattern attached to a precise 1D translation rig as illustrated in Fig. 5.5b. When the pattern is moved to a different position by the rig, its location can be given from reading the rig. So are those centers. To make the computations of the light planes consistent with those of the scene planes, a local system on the rig is defined and its relationship with the camera system is computed, say $\boldsymbol{R}_0$ and $\boldsymbol{t}_0$.

Given a position of the planar pattern, each light plane will intersect a line on the pattern and produce many points $\boldsymbol{M}_i$ with lines $\boldsymbol{L}_i$. Their correspondence $\boldsymbol{m}_i$ can be extracted from the image. It is known that given four collinear points in the image and three correspondences in 3D space, the remaining one can be calculated by cross ratio. So if we select three known points on $\boldsymbol{L}_i$ and extract their correspondence in the image, the point $\boldsymbol{M}_i$ can be calculated by cross ratio.

In summary, the procedure for determining the light planes is given as follows:

1. Computing the centers of three aligned circles on the planar pattern, given a position of the pattern;
2. Extracting the corresponding pixels of the three centers in the image and approximating a line;

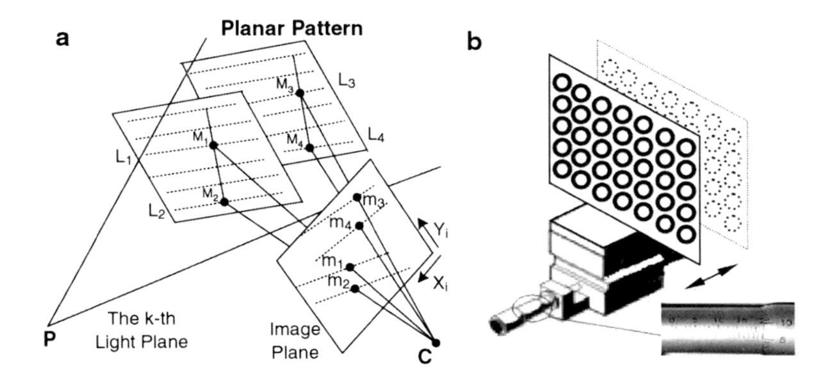

Fig. 5.5 A scheme for determining the light planes. **a** A number of points are produced by moving the planar pattern. **b** A 1D translation rig: the position of planar pattern can be read from its dial

3. In the image, extracting the intersecting line of the light plane and the planar pattern, then calculating the intersecting point of the above two lines;
4. Calculating the corresponding 3D point using cross ratio invariant given the three 3D points and four image points;
5. Moving the pattern to other positions to obtain other points on the light plane. Solving the equation for the light plane using three or more points on it;
6. Transforming the plane vector using the transformation $\begin{bmatrix} \boldsymbol{R}_0 & \boldsymbol{t}_0 \\ \boldsymbol{0}^{\mathrm{T}} & 1 \end{bmatrix}^{-\mathrm{T}}$ followed by $\begin{bmatrix} \boldsymbol{R}_1 & \boldsymbol{t}_1 \\ \boldsymbol{0}^{\mathrm{T}} & 1 \end{bmatrix}^{-\mathrm{T}}$, then obtaining the equation for the light plane expressed in the world system. Similar steps are implemented for other light planes. Once the light planes are calibrated, the transformation matrix $\boldsymbol{T}_k$ can be easily computed according to the formulas (5.1)–(5.5).

In practice, due to noise and computational errors, the points $\boldsymbol{M}_i$ in the 4th step may not all be exactly coplanar. Thus, we use the least squares method with all the points to determine an optimal plane in the 5th step, i.e.

$$\min \sum_i (\boldsymbol{n}^{\mathrm{T}} \boldsymbol{M}_i + d)^2 \tag{5.13}$$

Then we project these points onto the optimal plane and take the projection points as the estimations for them.

III. Discussions In this section, two different ways are talked about for determining the involved planes in the structured light system. One is based on the homographic transformation and the other is mainly based on cross ratio. In the latter, a 1-D mechanically controlled platform is needed. Its accuracy has a heavily impact on the calibration results. In addition, the calculation of cross ratio is sensitive to noise which requires elaborate treatment. In our experiments, a large number of points are used to increase the accuracy.

5.3.2 Determining the On-Line Homography

We are here aimed at uncalibrated reconstruction. In this system, the intrinsic and extrinsic parameters of the camera are allowed to be changed to "see" just the scene illuminated by the projector and obtain high quality image. Therefore, the Homography $\boldsymbol{H}_k$ between the image and light plane will be changed for different tasks and need to be determined dynamically.

There are two kinds of planes in the system: the two scene planes and two sets of light planes cast from the projector. They are non-collinear and intersect at many 3D points. Some of these points will appear on the object's surface and some on the two scene planes. As we have determined the equations for all these planes, the positions of those points appearing on the scene planes can be calculated. Their corresponding projections can be extracted from the images. As the Homographic matrix can be estimated with four or more pairs of non-collinear points, this problem is easily solved by the following procedure:

1. Detecting four or more points on the scene planes that are not occluded by the object and calculating their coordinates;
2. Transforming these points using (5.8) to obtain a set of points expressed in $F_{2\Pi k}$;
3. Extracting their corresponding pixels from the image captured by the CCD camera;
4. Computing the Homographic matrix $\boldsymbol{H}_k$ using (2.11) or (2.13).
5. Repeating steps 1–4 for other light planes. The results can be refined by a bundle adjustment algorithm after obtaining the initial values.

5.3.3 Euclidean 3D Reconstruction

In summary, the implementation of the structured light system for 3D reconstruction consists of two steps: static calibration and Euclidean reconstruction without knowing the camera's parameters. Assuming that the static calibration has been conducted as in Sect. 5.3.1, the procedure of the whole reconstruction process will consist of the following steps:

1. The object to be reconstructed is placed in front of the scene planes and illuminated by the projector. Based on the scene properties, adjust the camera's parameters including intrinsic and extrinsic parameters as needed, to enable the sensor to fully capture the illuminated scene. Then acquiring the image;
2. For each light plane, determining the image-to-plane Homographic matrix as in Sect. 5.3.2;
3. Computing the image-to-world transformation using the formula of (5.12);
4. For the points on the object's surface, performing Euclidean reconstruction using equation (5.11);

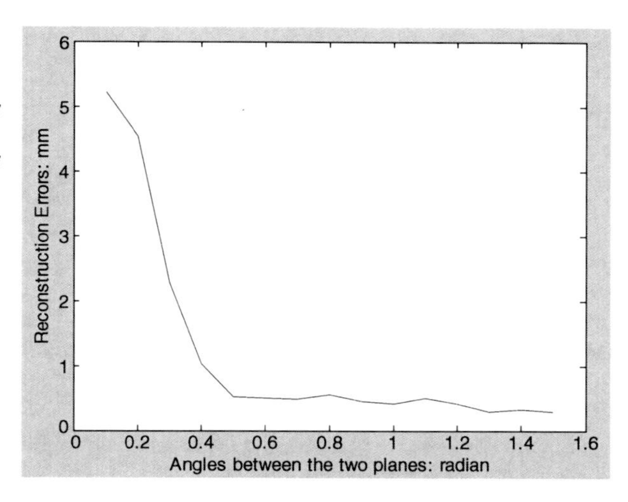

Fig. 5.6 Effects of the acute angle subtended by the two planes: *the smaller the angle, the closer the available points for the Homography to collinear, the larger the errors in 3D reconstruction will be*

5. Repeating steps 2–4 to estimate the object's points on other light planes, then we should obtain a cloud of 3D points for the concerned object.

5.3.4 Configuration of the Two Scene Planes

Theoretically, the system has no particular requirements on the configuration of the two scene planes as long as they are two different planes. But numerically, when the acute angle subtended by them is too small, the condition number of the coefficient matrix for the Homographic matrix becomes very large which will lead to an inaccurate estimation. And what's more, real data are generally noise-infected, it would render the system sensitive to noise and decrease the accuracy of reconstructed results in this case. So it is necessary to do a simulation to testify the effects of different angles to our system and optimize the configuration of our system thereafter. Herein, we ranged the angle from $\pi/30$ to $\pi/2$ with 15 intervals. For each interval, 100 trials were run and random noise with zero mean and constant derivation ($\sigma = 0.5$ pixel) was added to the simulated data.

Figure 5.6 showed the average result as a function of the angles. From these results, we can reach the conclusion that when the acute angle changes from very small value, say close to zero degree, to right angle, the reconstruction error drops drastically. From a geometrical point of view, the smaller the angle, the closer the available points for the Homography to collinear. In the extreme case, all the points will merge into a line when the angle is zero which would make it impossible to solve for the Homographic matrix. This indicates that two planes with perpendicular relationship will generally make the condition number of coefficient matrix small, which is much desired here.

5.3.5 Computational Complexity Study

In this section, we give an analysis on the computational complexity of the suggested method as compared with the traditional triangulation method, taking two cases into account intuitively yet reasonably. The result shows that the former is computationally more efficient in the 3D reconstruction than the latter.

I. Dealing with Similar Tasks When dealing with tasks of similar nature, it is not necessary to adjust the system from task to task after it has been properly set up. That is to say, the image-to-world transformation will be a constant matrix once it is determined. From (5.11), it can be seen that to reconstruct a world point, 12 multiplications are needed to determine the homogenous coordinates, followed by 3 divisions to convert them to Euclidean coordinates. In total, 15 multiplications are required if we assume division has the same complexity as a multiplication and ignoring the plus/minus operations.

For the traditional triangulation method, the computation of a world point involves in solving the following matrix equation:

$$\begin{bmatrix} P_{11} - P_{31}m_{i1} & P_{12} - P_{32}m_{i1} & P_{13} - P_{33}m_{i1} \\ P_{21} - P_{31}m_{i2} & P_{22} - P_{32}m_{i2} & P_{23} - P_{33}m_{i2} \\ a_k & b_k & c_k \end{bmatrix} \begin{bmatrix} M_{i1} \\ M_{i2} \\ M_{i3} \end{bmatrix} = \begin{bmatrix} m_{i1}P_{34} - P_{14} \\ m_{i2}P_{34} - P_{24} \\ -d_k \end{bmatrix} \tag{5.14}$$

where P_{ij} is the ij-th element of camera projection matrix $\boldsymbol{P}$ and $(m_{i1} \quad m_{i2})^{\mathrm{T}}$is an image pixel of the k-th light plane, a_k, b_k, c_k and d_k are its coefficients.

There are many methods for solving (5.14), including Gaussian Elimination, LU decomposition and Jacobi iteration method. Among these methods, Gaussian Elimination is the most economical whose computational complexity is

$$MD = \frac{n^3}{3} + n^2 - \frac{n}{3} \tag{5.15}$$

where n represents the number of equations.

From (5.15), 17 multiplications are needed with 8 extra multiplications in the first two rows to solve (5.14). Therefore, 25 operations in total are required.

From the above analysis, it is seen that the traditional triangulation method is computationally more costly than our method for the reconstruction, with at least 10 more multiplication operations needed to reconstruct a 3D point. This will introduce significant computational burden when dealing with large numbers of points for reconstructing complex or large objects.

Here, we have conducted a simulation experiment to see the difference in the computation time using different methods. The simulation was performed on a PC (Pentium 4, 2.8 GHz). In the test, the number of points to be reconstructed varied from 2,000 to 10,000. For each increment of 2,000, 100 tests were performed using both of our methods and the traditional triangulation method. The CPU time taken in the process of 3D reconstruction was recorded for each trial as in Table 5.1, where

Table 5.1 Simulation of the computational costs

Number of points	Triangulation method	Our method	Reduction in time
2,000	0.1540	0.0999	0.054
4,000	0.3449	0.2367	0.108
6,000	0.5809	0.4147	0.166
8,000	0.8484	0.6303	0.218
10,000	1.1755	0.8880	0.288

the second and third rows showed the average results (the time taken in second). From these results, we can see that the reduction in computation time is significant when the number of points becomes larger.

II. Dealing with Highly Diversified Tasks In such cases, online adjustments of the CCD camera such as auto-focus or unconstrained motion may frequently become necessary depending on the requirements of different tasks. As such, dynamic calibration of the image-to-world transformation must be performed during run time.

Here, the non-homogeneous solution with 4 pairs of points for the Homography is used. From (5.15), 232 multiplications with 16 extra multiplications are required to solve (2.11). Therefore, 248 operations in total are required to compute the Homographic matrix.

From (5.12), we know that 84 multiplications are needed to compute the online image-to-world transformation. Therefore, $248 + 84 = 332$ operations are required for each light stripe plane. Besides this, no calibration objects or any other operations are required for recalibration of our system. Such a recalibration process is very simple and efficient.

On the other hand, recalibration of CCD camera (such as intrinsic and extrinsic parameters, etc) is required if the traditional triangulation method is used. As a result, it is computationally much more costly to use a traditional method (such as direct linear transformation method) than our method, especially when frequent recalibration is necessary. Furthermore, the need of a delicate calibration object is too inconvenient for practical applications in those methods.

Based on the above two cases, we conclude that the image-to-world transformation is computationally more efficient than the triangulation method for 3D reconstruction. The improvement is remarkable especially when hundreds of thousands of points needed to be reconstructed for a large object or generating surrounding maps.

5.3.6 Reconstruction Examples

We used the structured light system constructed in Chap. 4 in the test of 3D reconstruction. Two areas each with $400 \times 500\,\text{mm}^2$ in size were selected acting as the two scene planes from the adjacent walls in our lab. According to Sect. 5.3.4, this improves the accuracy of the system since the two planes can be considered as perpendicular to each other with high precision. To determine their equations, two

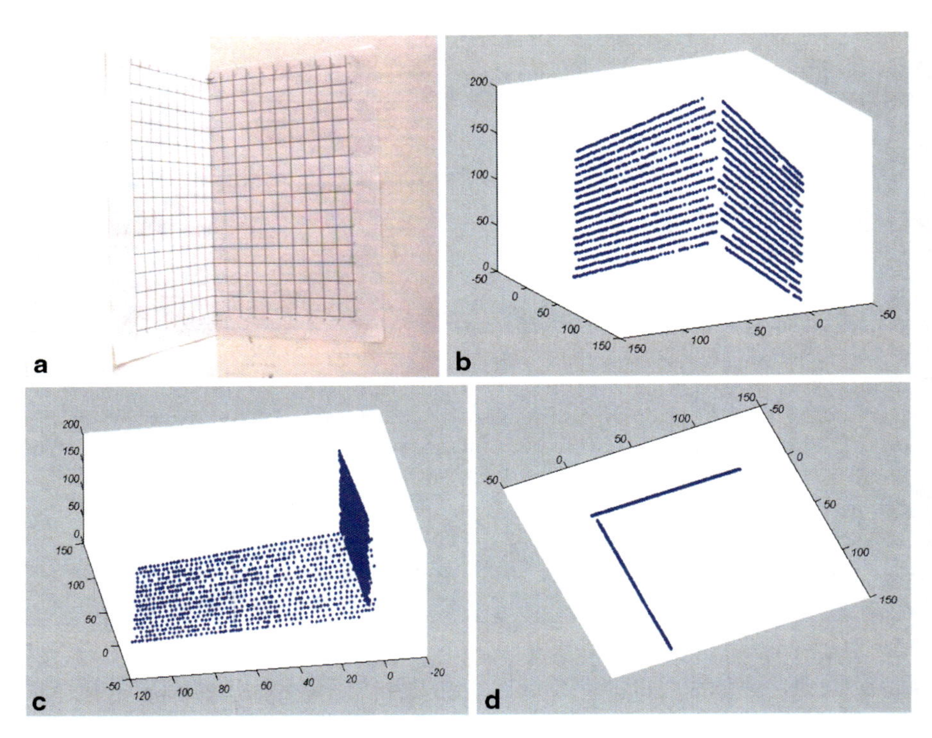

Fig. 5.7 Reconstructed corner walls. **a** Planar pattern used for calibration of the walls. **b** Reconstructed point clouds: a frontal view. **c** Reconstructed point cloud: a side view. **d** Reconstructed results: a top view

planar patterns containing many evenly distributed grids were attached as shown in Fig. 5.7a. The distance from the projector to the wall was around 1,000 mm to keep a reasonable field of view and illumination. The equations of the light planes and the scene planes were computed using the approaches in Sect. 5.3.1. The fitting errors, defined as the distances of the points from the fitted plane, were used to evaluate the accuracy of the light planes. Our experiments showed that the average fitting errors for the light planes could be very smaller, say less than 0.1 mm when 30 points were used to approximate each plane.

As our system is modeled by homographic transformation rather than the triangulation method, there is no requirement of particular baseline. During the reconstruction, the CCD camera was allowed to freely adjust (via motions, change of focus, etc.) according to the requirements of different tasks. To simulate the dynamic environment with moving objects, the camera and the objects were held by hand in image acquisition and subsequent 3D reconstruction. We performed the experiments for different types of objects to test our method. A typical and detailed evaluation of the relative errors was done on these experiments. It should be note that the grid pattern attached on the walls was used for static calibration, but not used in the subsequent reconstructions.

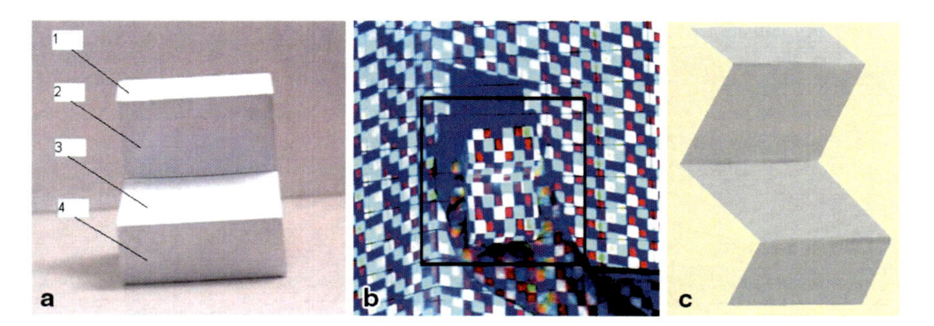

Fig. 5.8 Another experiment on a workpiece. **a** The workpiece used for the experiment. **b** The workpiece illuminated by the structured light pattern. **c** Reconstructed CAD model

Table 5.2 Fitting errors for surfaces of the workpiece (mm)

Index of surface	Max	Min	Mean
1	1.1377	2.9248e-005	0.4853
2	0.7054	0.0076	0.3153
3	0.8233	0.0011	0.2680
4	0.7628	7.2675e-004	0.3885

In the first example, we reconstructed the two adjacent walls. Figure 5.7b–d gave the frontal view, a side view and a top view of the reconstructed walls, where good coplanarity of the computed space points on each plane of the walls was observed. From these figures, we can see that the two reconstructed planes have no outlying points and form a nearly right angle. Here we gave the numerical evaluations on the coplanarity and orthogonality of the two planes.

Coplanarity of the Planes We fitted the two planes using these points and estimated the average values of the residual distances of the points to their planes. The value of one plane was 0.0533 mm, and the other was 0.0481 mm.

Orthogonality Relationship The angle subtended by the fitted planes was calculated to be 90.0944° against the true value of 90°, giving an error of 0.0944° or 0.1% in the relative error.

In the second example, we used a metal workpiece (Fig. 5.8a) to test the vision system. Here, we considered four surfaces labelled as 1, 2, 3 and 4. The workpiece was held by hand and illuminated by the projector as shown in Fig. 5.8b (encapsulated part by bold black lines) and the surfaces were reconstructed using the proposed procedure. Their fitting errors and metric sizes were measured and given in Tables 5.2 and 5.3 respectively. From these results, we can see that the differences of the reconstructed points from the fitted surface are quite small. The reconstructed CAD model was shown in Fig. 5.8c. To demonstrate the correct spatial relationships between different surfaces, the angular errors between them were evaluated and the results were given in Table 5.4, where nearly perpendicular relationships between

Table 5.3 Evaluation of measurements of the metric sizes

Index of surface	Real value (mm)	Meas. value (mm)	Relative error (%)
1	25	25.2075	0.83
2	35	34.5719	1.22
3	45	45.6799	1.51
4	25	25.4290	1.72

Table 5.4 Evaluation of measurements for the angles

Angle between planes	Actual value θ_0 (°)	Measured value θ (°)	Relative error $\Delta\theta/\theta_0$ (%)
1, 2	90	91.3545	1.51
2, 3	90	90.4879	0.54
3, 4	90	91.0950	1.22

consecutive surfaces were observed as expected. As our system is linear, the errors are mainly propagated from the possible errors in the static calibration and those from the image processing. In the current implementation, we do not make special efforts in the image processing or the initial calibration of the projector, as we aim mainly at verifying the validity of our method.

5.4 One-Known-Plane Based Method

In the previous section, two scene planes are required for automatic calibration. This may be fulfilled in a variety of environments, such as corner of a building or an office. However, the requirement of one scene plane may be more reasonable and convenient in some applications. For example, in robot navigation, we can assume the ground plane or the road surface is always existed and fixed. Besides, more and more applications are found in a desktop vision system (DVS). Exploring relevant problems in this case seems to be considerably meaningful. Therefore, we will present relevant investigations on this issue in what follows.

5.4.1 Calibration Tasks

This scenario is sketched in Fig. 5.9, in which there is one known plane in the scene and the object to be reconstructed is placed in front of it. The projector is also considered as a group of light planes and will produce many line segments or curves on the object's surface and the scene plane. The camera captures the illuminated scene in an image. Some of the intrinsic and extrinsic parameters of the camera can be changed to adapt to different objects when working. Here, the key issue is how to determine the image-to-world transformation automatically. Different from the first case with two scene planes, the image-to-light-plane Homography cannot be

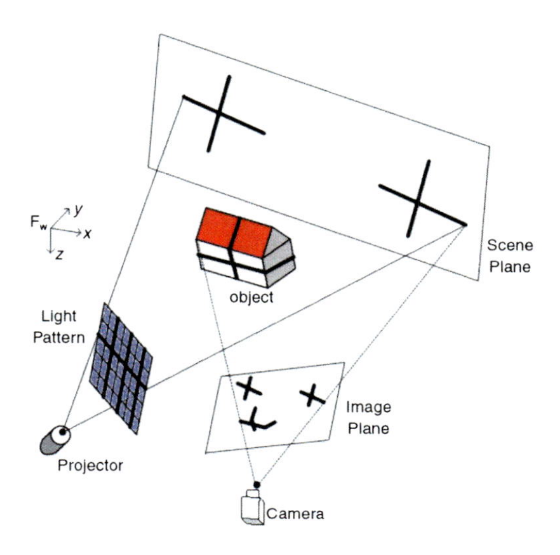

Fig. 5.9 A sketch of the structured light system

computed directly since only one scene plane is involved and only collinear points are obtained for each light plane. To overcome this difficulty, we firstly drive a formula of generic Homography, from which we can calculate the image-to-scene-plane Homography and then the image-to-light plane Homography. The former is used to estimate the variable parameters of the system while the latter facilitates the estimation of the image-to-world transformation.

Based on the above analysis, we can see the intrinsic and extrinsic parameters of the camera would be required in such system. So the first task is concerning with the calibration of intrinsic parameters of the camera and equations of those light planes as well as the scene plane. This belongs to static calibration, and can be easily completely by the algorithms in Chap. 3 and the procedures in Sect. 5.3.1. The second one deals with the calibration of the extrinsic parameters and possibly variable intrinsic parameters of the camera with the results of static calibration. Then the image-to-light plane Homography and subsequently the image-to-world transformation can be computed. This step belongs to dynamic calibration. Again, the static calibration needs to be performed only once while the dynamic calibration is frequently required whenever the vision system is reconfigured. We will talk about some details for the dynamic calibration in the next section.

5.4.2 Generic Homography

The Homography is a collinear mapping between two planes. In general, it should not be restricted to a specific kind of planes. From (5.10), for a point $\widetilde{\boldsymbol{M}}_p$ on an arbitrary plane, we will have

$$\widetilde{\boldsymbol{M}}_p = \alpha \boldsymbol{T}_p^{-1} \boldsymbol{S}^{\mathrm{T}} \widetilde{\boldsymbol{m}}_p \tag{5.16}$$

Substituting (5.16) into (2.3), we obtain

$$\tilde{\boldsymbol{m}} = \alpha\beta \boldsymbol{K}_c[\boldsymbol{R}_c \quad \boldsymbol{t}_c]\boldsymbol{T}_p^{-1}\boldsymbol{S}^{\mathrm{T}}\tilde{\boldsymbol{m}}_p \tag{5.17}$$

Combining (2.9) and (5.17), we have

$$\lambda \boldsymbol{H}_p = \boldsymbol{K}_c[\boldsymbol{R}_c \quad \boldsymbol{t}_c]\boldsymbol{T}_p^{-1}\boldsymbol{S}^{\mathrm{T}} \tag{5.18}$$

where λ is a nonzero scale factor.

The above formula is an explicit expression of the plane-to-image Homography. It is a 3×3 matrix and can be determined up to a scale factor. It transforms a point in a plane to a pixel in the camera image. Here, the plane can represent either a scene plane or a light plane.

Discussion If the world frame coincides with the local frame of the plane and $\overline{\boldsymbol{M}}$ is the origin of the coordinate frame, then $\boldsymbol{R}_p = \boldsymbol{I}$. From the expression of $\boldsymbol{T}_p$ in (5.5), we have $\boldsymbol{T}_p^{-1} = \begin{bmatrix} \boldsymbol{I} & \boldsymbol{0} \\ \boldsymbol{0}^{\mathrm{T}} & 1 \end{bmatrix}$. Then from the right side of (5.18), we obtain

$$\boldsymbol{K}_c\begin{bmatrix} \boldsymbol{c}_1 & \boldsymbol{c}_2 & \boldsymbol{c}_3 & \boldsymbol{t}_c \end{bmatrix}\begin{bmatrix} \boldsymbol{I} & \boldsymbol{0} \\ \boldsymbol{0}^{\mathrm{T}} & 1 \end{bmatrix}\begin{bmatrix} 1 & 0 & 0 \\ 0 & 1 & 0 \\ 0 & 0 & 0 \\ 0 & 0 & 1 \end{bmatrix} = \boldsymbol{K}_c\begin{bmatrix} \boldsymbol{c}_1 & \boldsymbol{c}_2 & \boldsymbol{t}_c \end{bmatrix}.$$

where $\boldsymbol{c}_1$ and $\boldsymbol{c}_2$ denote the first two columns of matrix $\boldsymbol{R}_c$. It is seen that this result coincides with that of Zhang's [34]. In fact, Zhang's result is a special case, that is when $\boldsymbol{R}_p$ is an identity matrix and $\overline{\boldsymbol{M}}$ is a zero vector. In the formula of (5.18), the values for $\boldsymbol{R}_p$ and $\overline{\boldsymbol{M}}$ can be arbitrary. This is necessary since multiple planes including the light planes and the scene plane are involved in this system. We call it a generic Homography as it encodes them all in a unique formula.

5.4.3 Dynamic Calibration

In this system, those feature points on the scene plane can be calculated by three relevant planes as their equations are calibrated statically. Hence, the Homographic matrix between the scene plane and its image can be calculated by a point-to-point method using (2.11) or (2.13) with four or more illuminated point correspondences. Let $\boldsymbol{H}_s$ denote the computed Homography, where the subscript 's' represents the scene plane relevant items. In what follows, it is this matrix that is used for dynamic calibration. Here, two different cases are considered with the Homography. The first case deals with the extrinsic parameters assuming all the intrinsic parameters are known. In the second case, the focal lengths in both pixel dimensions as well as the extrinsic parameters of the camera can be variable.

I. Fixed Intrinsic Parameters In this case, we assume that the intrinsic parameters of the camera is known and fixed while the pose parameters of the camera can be variable. So the major task is to estimate the extrinsic parameters. From (5.18), we have

$$\lambda \boldsymbol{A} = \boldsymbol{B} \tag{5.19}$$

where $\boldsymbol{A} = \boldsymbol{K}_c^{-1}\boldsymbol{H}_s$ and $\boldsymbol{B} = [\boldsymbol{R}_c \quad \boldsymbol{t}_c]\boldsymbol{T}_s^{-1}\boldsymbol{S}^{\mathrm{T}}$.

We will firstly discuss how to determine the scale factor λ and then the rotation matrix $\boldsymbol{R}_c$ and the translation vector $\boldsymbol{t}_c$.

Proposition 5.1 *Let* $\boldsymbol{B} = [\boldsymbol{R}_c \quad \boldsymbol{t}_c]\boldsymbol{T}_s^{-1}\boldsymbol{S}^{\mathrm{T}}$. *The first two columns in the matrix* $\boldsymbol{B}$ *have a unit norm each.*

Proof From (5.5), the inverse matrix of $\boldsymbol{T}_s$ is $\boldsymbol{T}_s^{-1} = \begin{bmatrix} \boldsymbol{R}_s^{\mathrm{T}} & \overline{\boldsymbol{M}} \\ \boldsymbol{0}^{\mathrm{T}} & 1 \end{bmatrix}$ since $\boldsymbol{T}_s \cdot \boldsymbol{T}_s^{-1} = \begin{bmatrix} \boldsymbol{I} & \boldsymbol{0} \\ \boldsymbol{0}^{\mathrm{T}} & 1 \end{bmatrix}$.

Let $\boldsymbol{R}_c = \begin{bmatrix} \boldsymbol{r}_1 \\ \boldsymbol{r}_2 \\ \boldsymbol{r}_3 \end{bmatrix}$ and $\boldsymbol{R}_s = \begin{bmatrix} \boldsymbol{r}_{1s} \\ \boldsymbol{r}_{2s} \\ \boldsymbol{r}_{3s} \end{bmatrix}$, where $\boldsymbol{r}_i$ and $\boldsymbol{r}_{is}$ are the i-th row of $\boldsymbol{R}_c$ and $\boldsymbol{R}_s$ respectively. Let $\boldsymbol{t}_c = [t_1 \quad t_2 \quad t_3]^{\mathrm{T}}$.

Then

$$\begin{aligned} \boldsymbol{B} &= \begin{bmatrix} \boldsymbol{r}_1 & t_1 \\ \boldsymbol{r}_2 & t_2 \\ \boldsymbol{r}_3 & t_3 \end{bmatrix} \cdot \begin{bmatrix} \boldsymbol{r}_{1s}^{\mathrm{T}} & \boldsymbol{r}_{2s}^{\mathrm{T}} & \boldsymbol{r}_{3s}^{\mathrm{T}} & \overline{\boldsymbol{M}} \\ 0 & 0 & 0 & 1 \end{bmatrix} \begin{bmatrix} 1 & 0 & 0 \\ 0 & 1 & 0 \\ 0 & 0 & 0 \\ 0 & 0 & 1 \end{bmatrix} \\ &= \begin{bmatrix} \boldsymbol{r}_1\boldsymbol{r}_{1s}^{\mathrm{T}} & \boldsymbol{r}_1\boldsymbol{r}_{2s}^{\mathrm{T}} & \boldsymbol{r}_1\overline{\boldsymbol{M}} + t_1 \\ \boldsymbol{r}_2\boldsymbol{r}_{1s}^{\mathrm{T}} & \boldsymbol{r}_2\boldsymbol{r}_{2s}^{\mathrm{T}} & \boldsymbol{r}_2\overline{\boldsymbol{M}} + t_2 \\ \boldsymbol{r}_3\boldsymbol{r}_{1s}^{\mathrm{T}} & \boldsymbol{r}_3\boldsymbol{r}_{2s}^{\mathrm{T}} & \boldsymbol{r}_3\overline{\boldsymbol{M}} + t_3 \end{bmatrix} \end{aligned} \tag{5.20}$$

Let the first column of matrix $\boldsymbol{B}$ be $\boldsymbol{b}_1$. Then we have

$$\boldsymbol{b}_1 = \begin{bmatrix} \boldsymbol{r}_1\boldsymbol{r}_{1s}^{\mathrm{T}} \\ \boldsymbol{r}_2\boldsymbol{r}_{1s}^{\mathrm{T}} \\ \boldsymbol{r}_3\boldsymbol{r}_{1s}^{\mathrm{T}} \end{bmatrix} = \begin{bmatrix} \boldsymbol{r}_1 \\ \boldsymbol{r}_2 \\ \boldsymbol{r}_3 \end{bmatrix} \cdot \boldsymbol{r}_{1s}^{\mathrm{T}} = \boldsymbol{R}_c \cdot \boldsymbol{r}_{1s}^{\mathrm{T}} \tag{5.21}$$

This gives

$$\|\boldsymbol{b}_1\| = \sqrt{(\boldsymbol{R}_c \cdot \boldsymbol{r}_{1s}^{\mathrm{T}})^{\mathrm{T}}(\boldsymbol{R}_c \cdot \boldsymbol{r}_{1s}^{\mathrm{T}})} = \sqrt{\boldsymbol{r}_{1s}\boldsymbol{R}_c{}^{\mathrm{T}}\boldsymbol{R}_c\boldsymbol{r}_{1s}^{\mathrm{T}}} = \sqrt{\boldsymbol{r}_{1s}\boldsymbol{r}_{1s}^{\mathrm{T}}} = 1 \tag{5.22}$$

The similar result can be obtained for the second column of matrix $\boldsymbol{B}$.

So the first two columns have a unit norm each. □

Let $\boldsymbol{a}_1$ be the first column of matrix $\boldsymbol{A}$. According to proposition 5.1, the scale factor can be calculated by

$$\lambda = \frac{1}{\|\boldsymbol{a}_1\|} \tag{5.23}$$

The sign of the scale factor should make both sides of (2.9) consistent. With i pairs of points, if

$$\lambda \sum_i \boldsymbol{m}_{c,i}^{\mathrm{T}} \boldsymbol{H} \boldsymbol{m}_{p,i} < 0 \tag{5.24}$$

its sign should be reversed. One pair of projection points is sufficient to determine the sign. In practice, we use two or more pairs to increase its reliability as in (5.24).

Scaling the matrix $\boldsymbol{A}$ by the scale factor, we have

$$\widehat{\boldsymbol{A}} = \boldsymbol{B} \tag{5.25}$$

where: $\widehat{\boldsymbol{A}} = \lambda \boldsymbol{A}$.

Now, we will discuss the way for resolving $\boldsymbol{R}_c$ and $\boldsymbol{t}_c$.

Considering the first row in both sides of (5.25), we can construct the equations for $\boldsymbol{r}_1$,

$$\begin{cases} \boldsymbol{r}_1 \boldsymbol{r}_{1s}^{\mathrm{T}} = \widehat{\boldsymbol{A}}_{11} \\ \boldsymbol{r}_1 \boldsymbol{r}_{2s}^{\mathrm{T}} = \widehat{\boldsymbol{A}}_{12} \\ \|\boldsymbol{r}_1\|_2 = 1 \end{cases} \tag{5.26}$$

where $\widehat{\boldsymbol{A}}_{ij}$ is the ij-th element of the matrix $\widehat{\boldsymbol{A}}$.

There are three equations in (5.26) for the three unknowns. So we can obtain the solution for $\boldsymbol{r}_1$. Similarly, $\boldsymbol{r}_2$ can be solved by considering their second row. According to the property of rotation matrix, $\boldsymbol{r}_3$ can be calculated as a cross product of the first two rows, i.e. $\boldsymbol{r}_3 = \boldsymbol{r}_1 \times \boldsymbol{r}_2$. Then we have the rotation matrix $\boldsymbol{R}_c$. Although there exist multiple solutions, only one is correct and the others can be eliminated by back-projection method.

Considering the third column of (5.25), we get the following equations for the translation vector $\boldsymbol{t}_c$:

$$\begin{cases} t_1 = \widehat{\boldsymbol{A}}_{13} - \boldsymbol{r}_1 \overline{\boldsymbol{M}} \\ t_2 = \widehat{\boldsymbol{A}}_{23} - \boldsymbol{r}_2 \overline{\boldsymbol{M}} \\ t_3 = \widehat{\boldsymbol{A}}_{33} - \boldsymbol{r}_3 \overline{\boldsymbol{M}} \end{cases} \tag{5.27}$$

We have thus solved the rotation matrix and translation vector for the camera. Once $\boldsymbol{R}_c$ and $\boldsymbol{t}_c$ have been obtained, the image-to-light plane Homography can be estimated using (5.18). Then the calibration task is finished with the computation of the image-to-world transformation using (5.12).

II. With Variable Focal Lengths In this case, the simplified camera matrix is used which means there are two variable parameters in $\boldsymbol{K}_c$. So the major task is to calibrate the variables as well as the extrinsic parameters.

Considering the image-to-scene plane Homography from (5.18), we have

$$\lambda \boldsymbol{H}_s = \begin{bmatrix} f_u \boldsymbol{r}_1 \boldsymbol{r}_{1s}^{\mathrm{T}} & f_u \boldsymbol{r}_1 \boldsymbol{r}_{2s}^{\mathrm{T}} & f_u(\boldsymbol{r}_1 \overline{\boldsymbol{M}} + t_1) \\ f_v \boldsymbol{r}_2 \boldsymbol{r}_{1s}^{\mathrm{T}} & f_v \boldsymbol{r}_2 \boldsymbol{r}_{2s}^{\mathrm{T}} & f_v(\boldsymbol{r}_2 \overline{\boldsymbol{M}} + t_2) \\ \boldsymbol{r}_3 \boldsymbol{r}_{1s}^{\mathrm{T}} & \boldsymbol{r}_3 \boldsymbol{r}_{2s}^{\mathrm{T}} & \boldsymbol{r}_3 \overline{\boldsymbol{M}} + t_3 \end{bmatrix} \tag{5.28}$$

From the first two columns in (5.28), we obtain

$$\boldsymbol{R}_c \boldsymbol{r}_{1s}^{\mathrm{T}} = \begin{bmatrix} \left(\frac{\lambda}{fu}\right) \boldsymbol{H}_{11} \\ \left(\frac{\lambda}{f_v}\right) \boldsymbol{H}_{21} \\ \lambda \boldsymbol{H}_{31} \end{bmatrix} \tag{5.29}$$

and

$$\boldsymbol{R}_c \boldsymbol{r}_{2s}^{\mathrm{T}} = \begin{bmatrix} \left(\frac{\lambda}{fu}\right) \boldsymbol{H}_{12} \\ \left(\frac{\lambda}{f_v}\right) \boldsymbol{H}_{22} \\ \lambda \boldsymbol{H}_{32} \end{bmatrix} \tag{5.30}$$

where $\boldsymbol{H}_{ij}$ means the element on the i-th row and j-th column of the matrix $\boldsymbol{H}_s$.

From (5.29) and (5.30), we will have the following three equations:

$$\left(\left(\frac{\lambda}{f_u}\right) \boldsymbol{H}_{11}\right)^2 + \left(\left(\frac{\lambda}{f_v}\right) \boldsymbol{H}_{21}\right)^2 + (\lambda \boldsymbol{H}_{31})^2 = 1 \tag{5.31}$$

$$\left(\left(\frac{\lambda}{f_u}\right) \boldsymbol{H}_{12}\right)^2 + \left(\left(\frac{\lambda}{f_v}\right) \boldsymbol{H}_{22}\right)^2 + (\lambda \boldsymbol{H}_{32})^2 = 1 \tag{5.32}$$

$$\left(\frac{\lambda}{f_u}\right)^2 \boldsymbol{H}_{11} \boldsymbol{H}_{12} + \left(\frac{\lambda}{f_v}\right)^2 \boldsymbol{H}_{21} \mathbf{H}_{22} + \lambda^2 \boldsymbol{H}_{31} \boldsymbol{H}_{32} = 0 \tag{5.33}$$

In (5.31), (5.32) and (5.33), there are three unknowns, i.e. f_u, f_v and λ. From these equations, the solution can be given as

$$f_u = \sqrt{\frac{(\boldsymbol{H}_{11} \boldsymbol{H}_{22} - \boldsymbol{H}_{12} \boldsymbol{H}_{21})(\boldsymbol{H}_{11} \boldsymbol{H}_{21} + \boldsymbol{H}_{22} \boldsymbol{H}_{12})}{(\boldsymbol{H}_{21} \boldsymbol{H}_{32} - \boldsymbol{H}_{22} \boldsymbol{H}_{31})(\boldsymbol{H}_{21} \boldsymbol{H}_{31} + \boldsymbol{H}_{22} \boldsymbol{H}_{32})}}, \tag{5.34}$$

$$f_v = \sqrt{\frac{(\boldsymbol{H}_{11}\boldsymbol{H}_{22} - \boldsymbol{H}_{12}\boldsymbol{H}_{21})(\boldsymbol{H}_{11}\boldsymbol{H}_{21} + \boldsymbol{H}_{22}\boldsymbol{H}_{12})}{(\boldsymbol{H}_{12}\boldsymbol{H}_{31} - \boldsymbol{H}_{11}\boldsymbol{H}_{32})(\boldsymbol{H}_{11}\boldsymbol{H}_{31} + \boldsymbol{H}_{32}\boldsymbol{H}_{12})}} \tag{5.35}$$

and

$$\lambda = \pm\sqrt{\frac{f_u^2 + f_v^2}{f_u^2 f_v^2 \boldsymbol{H}_{31}^2 + f_u^2 \boldsymbol{H}_{21}^2 + f_v^2 \boldsymbol{H}_{11}^2}} \tag{5.36}$$

Since both f_u and f_v should be positive, unique solutions are obtained. As to λ, its sign can be determined by (5.24). Then the remaining problem becomes solving for the extrinsic parameters and similar process can be taken as in the first case.

Once all the variable parameters of the system are calibrated, the image-to-light-plane Homography can be computed according to formula (5.18). Subsequently, the Euclidean 3D Reconstruction is performed with the image-to-world transformation obtained by formula (5.12).

5.4.4 Reconstruction Procedure

Objectives: With the necessary static calibration done, estimating the image-to-light-plane Homography and image-to-world transformation online, then performing 3D Euclidean reconstruction.

Procedure:

Step 1: Given a set of point correspondences between the image and the scene plane, computing the image-to-scene-plane Homography by (2.11) or (2.13).
Step 2: Estimating the variable parameters using the Homography. If the focal lengths of the camera are fixed, obtaining the rotation matrix and translation vector with (5.26) and (5.27). Otherwise, calculating them according (5.34) and (5.34) first.
Step 3: Computing the image-to-light-plane Homography $\boldsymbol{H}_k$ according to (5.18).
Step 4: Estimating the image-to-world transformation $\boldsymbol{T}_{wk}$ with (5.12).
Step 5: Performing 3D reconstruction using (5.11).

5.4.5 Reconstruction Examples

The structured light system constructed previously was employed for the 3D reconstruction experiments in this section. In the following examples, a piece of planar area from the vertical wall in our lab played the role of the scene plane. The concerned objects were placed in front of the plane. Figure 5.11a shows a profile of the experiment scenario.

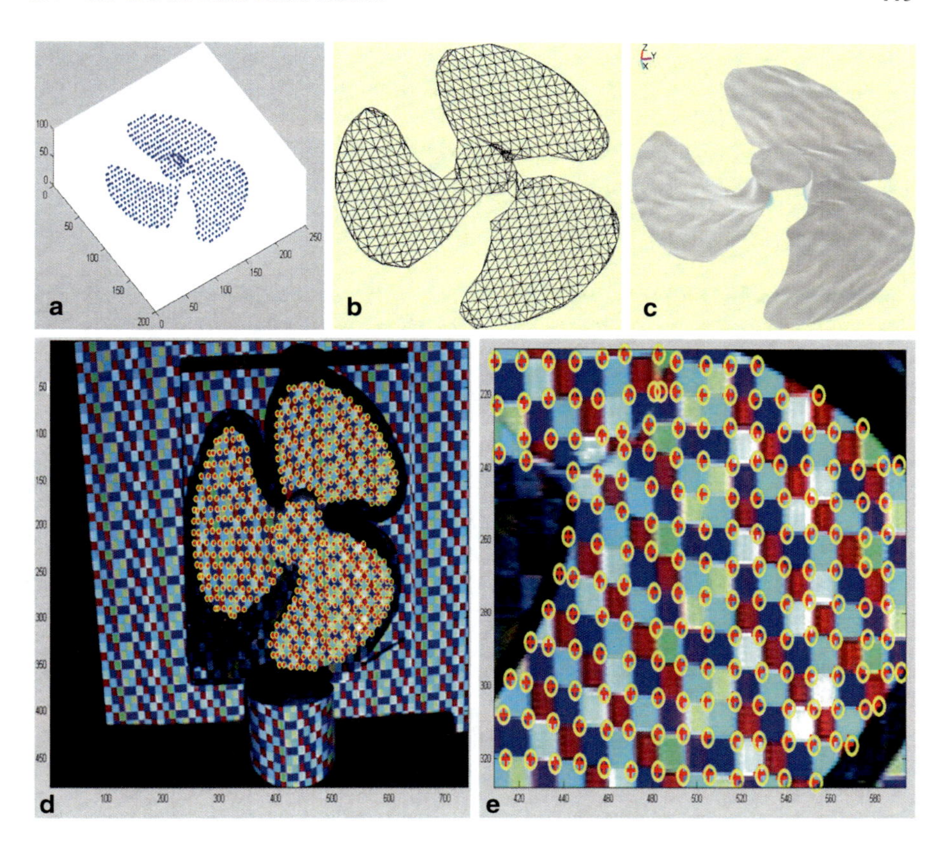

Fig. 5.10 Experiment on the fan model. **a** Reconstructed 3D point clouds. **b** Polygonized results of the points clouds. **c** A CAD model of the points clouds. **d** Original feature points and back-projected points. *Red* '+' represents original feature points while *yellow* '*o*' represents back-projecting points from reconstructed 3D points. They should coincide with each other theoretically. **e** Zoomed part of the image. It is obvious that they are very closer to each other as expected

We firstly tested on a fan model with freeform surface as shown in Fig. 5.10d. Totally 603 feature points from its surface were reconstructed using the suggested procedure. Figure 5.10a showed the reconstructed 3D point clouds of the model's surface. Figure 5.10b, c illustrated the polygonized results and the CAD model of the point clouds. Since no attempt has been made to obtain the ground truth, we do not know real values of the variable parameters of the system and the 3D point clouds. To quantitatively evaluate their accuracy, these point clouds were back-projected into the original image with the calibration results. Then the discrepancies between the real feature points and back-projected points were computed. In general, the more accurate the calibrated results, the smaller discrepancies we will have. Here, the minimum and maximum absolute discrepancies were found to be 0.0249 pixel and 2.1410 pixels respectively while the mean was 0.7595 pixel. For comparison, we

Table 5.5 Comparison of the residual 2D errors

Different methods		Max. abs. error (pixel)	Mean. abs. error (pixel)
Fofi's results	Linear method	50.084	18.428
	Iterative method	3.069	0.204
Our results	Linear method	2.1410	0.7595

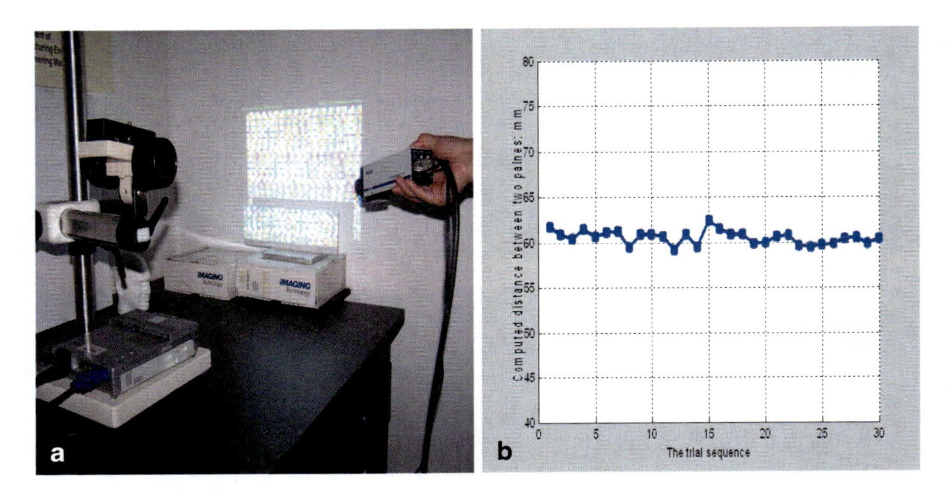

Fig. 5.11 An example of the demonstration system. **a** A profile of the system setup. **b** The measured distance between the two planes

listed our results together with those from Fofi [172] in Table 5.5. The improvement is observed in this table considering that only linear algorithm is involved in the suggested procedure. To visualize these results, we plotted the feature points and back-projected points in Fig. 5.10d, e illustrated the zoomed part of the image for graceful view from which we can see that the feature points and back-projected points are very close to each other. This indicates the high accuracy of the calibration results.

In the second example, we had developed a demonstration system in order to explore the applicability of the method in which a hand-held camera was used. In this example, the focal lengths of the camera were calibrated and kept fixed while its extrinsic parameters was arbitrarily moved by hand when working (Fig. 5.11a). Two parallel planes were reconstructed and their distance was measured automatically and consecutively. The results of 30 sequential trials were recorded and illustrated in Fig. 5.11b. Here, the mean of measured distance was 60.52 mm and the standard deviation was 0.76 mm. This result is acceptable considering the real distance between the two planes is set to be 60 mm.

5.5 Summary

We have talked about approaches for uncalibrated 3D reconstruction by using the image-to-world transformation via a color encoded structured light system. Here, the projector in the system is regarded as a collection of light planes rather than as a virtual camera and the plane-based Homography are extensively explored to make full use of those planes.

Two different cases are considered: In the first case, two known planes are assumed in the workspace, where the image-to-world transformation can be automatically estimated regardless of the intrinsic and extrinsic parameters of the camera. Hence, the focusing and unconstrained motions of the camera are allowed whenever it is necessary. In the second case, one known plane in the workspace is involved. Here, the scene-plane-to-image Homography is used for determining the variable intrinsic or extrinsic parameters of the camera. Then the image-to-light-plane Homography is obtained from the generic formula and so is the image-to-world transformation. In practice, the requirement of one known plane is certainly easier to be satisfied than that of two known planes, and promises more flexibilities and applications.

The computational complexity of the 3D reconstruction algorithm is analyzed and compared with the traditional triangulation method. We also show by simulated experiment that our system will be optimized when the two scene planes are perpendicular to each other. The acquisition speed of images is instantaneous as only one image is required for 3D reconstruction. Both numerical simulations and real data experiments are carried out to test the system. Quantitative evaluations of the relative errors demonstrate that the experimental results are accurate and reliable. As the system is linear, these errors are generally propagated from the residual errors during the image processing and initial calibration of the system components. With more sophisticated methods adopted in the image processing and initial calibration, higher accuracy can be expected with the proposed method. To simulate dynamic scenes with moving objects, the camera and the objects to be reconstructed are held by hand and moved freely. In a robotic application, a number of different tasks such as object recognition, object measurements or object acquisition by a manipulator will be involved and this technique provides a very good solution.

There are also some limitations with this method. The projection of color patterns favors color-neutral surface of the objects to be measured. When dealing with colorful surface, a frequency shifting happens to the color captured by the camera with respect to the color projected by the light pattern. Then the pattern decoding process will be complicated and unreliable. From an engineering point of view, this is acceptable as most industrial parts are textureless and monochromic. The assumption of flat scene planes will introduce errors if they are not perfectly planar. However, this can be quantified off-line.

Chapter 6
Catadioptric Vision System

Wide field of view of a vision system is desirable in many applications. The Catadioptric vision system, which combines reflective mirror and refractive lens, is an effective way to enlarge its viewing angle compared with the traditional system. In this chapter, we will present some recent development in the related research field. We firstly introduce the backgrounds of the Catadioptric vision system. Then three different systems are elaborated and they are: (1) panoramic stereoscopic system which cooperates a fisheye lens camera and a convex reflective mirror; (2) parabolic camera system which consists of a paraboloid-shaped mirror and a CCD camera with orthographic projection; and (3) hyperbolic camera system which combines a hyperboloid-shaped mirror and a CCD camera. All of these systems have two important characteristics: one is a single effective viewpoint and the other is wide field of view. The mathematical models of the imaging process are investigated and the calibration algorithms with different tools are explored.

6.1 Introduction

6.1.1 Wide Field-of-View System

In practice, it is highly desirable to endow a vision system with a wider field of view since a larger part of the scenery can be observed in a single shot. However, the traditional CCD camera usually has a limited view angle. For example, a camera with 1/3 inch sensor and 8 mm lens can only provide about 50° in the horizontal angle and 40° in vertical angle. So the vision system with this kind of cameras is easy to miss the concerned objects or their feature points in a dynamic environment. Furthermore, it is often difficult to obtain sufficient overlap between different images, which will lead to troublesome when performing the vision tasks. Of course, such problem can be generally solved if the view field of the system is enlarged.

There are many approaches to construct a vision system with large view field. We can roughly classify them into the following four categories. The first category is to

B. Zhang, Y. F. Li, *Automatic Calibration and Reconstruction for Active Vision Systems*, Intelligent Systems, Control and Automation: Science and Engineering,
DOI 10.1007/978-94-007-2654-3_6, © Springer Science+Business Media B.V. 2012

Fig. 6.1 Camera matrix constructed by Stanford Computer Graphics Laboratory

Fig. 6.2 The PTZ camera from http://en.wikipedia.org/wiki/Pan_tilt_zoom_camera

simultaneously use many cameras and combine their view fields together. Figure 6.1 shows two different arrangements of the camera matrix systems, in which each camera can adjust its exposure independently, allowing to recording high dynamic range environments. The main problem in this kind of vision system is how to synchronously control those cameras. And such system is so bulky that it is not appropriate for applications of mobile robotics. The use of a rotating camera to capture sequential images covering the whole concerned scene can be considered as the second category. A typical setup for this kind of system is a PTZ camera for visual surveillance, which can be moved from left to right, up and down, and zoom in and out as illustrated in Fig. 6.2. In these examples, the field of view is enlarged either by combining multiple cameras or by motions. However, the view angle does not be essentially changed for a single camera.

Fig. 6.3 The top left shows Nikon Coolpix digital camera and Nikon FC-E8 fish-eye lens; the top right gives Canon EOS-1Ds and Sigma 8 mm-f4-EX fish-eye lens

In nature, most species with lateral eye placement have an almost spherical view angle, for example the insect eyes. In the vision community, the large field of view can be achieved with pure dioptric elements. Here, this kind of system falls into the third category, i.e. a camera system with fisheye lens. The distinct advantage of such system is that it itself can provide extremely wide view angle. In general, there are two kinds of fisheye lenses, i.e. circular and full-frame fisheye lens. Figure 6.3 shows two vision systems with circular fisheye lens given in [190]. From their images below, we can see that the field of view is as large as about 180°. Hence, they can provide images of large areas of the surrounding scene with a single shot. The fundamental difference between a fisheye lens and the classical pin-hole lens is that the projection from 3D rays to 2D image in the former is intrinsically non perspective, which makes the classical imaging model invalid. As a result, it is difficult to give a closed-from model for the imaging process. In the literature, the Taylor expansion model, rational model or division model is frequently adopted for such vision system.

The camera-lens-mirror system is considered as in the last category, which composes with the traditional camera and curved mirror with special shape to enhance the sensor's field of view [191, 209]. Since both reflective (cataoptric) and refractive (dioptric) rays are involved, the system is also named as catadioptric camera system or omni-directional system. Figure 6.4 shows two typical systems. For examples, the horizontal and vertical view-angle can respectively reach as large as 360° and 115°

Fig. 6.4 Two examples of the catadioptric camera systems. **a** The 0-360 Panoramic Optic. **b** Picture from the Institute for Computer Graphics and Vision

for the system in Fig. 6.4a (from www.0-360.com). Compared to the traditional system with narrow field of view, there are several advantages for such systems: Firstly, the search for feature correspondences is easier since the corresponding points do not often disappear from the images; Secondly, a large field of view stabilizes the motion estimation algorithms; Lastly but not least, more information of the scene or interested objects can be reconstructed from fewer images. As a result, it offers great benefits to visual surveillance, teleconferencing, three-dimensional modeling of wide environment [187, 188] and robot navigation [189], etc.

Many different types of mirrors can be employed in such system, including planar mirrors, elliptical mirrors, parabolic and hyperbolic mirrors, etc. Accordingly, they can be categorized into parabolic camera system (combining a parabolic mirror with an orthographic camera) and hyperbolic camera system (a hyperbolic mirror with a perspective camera), etc. On the other hand, depending on whether or not all the incident rays will pass through a single point named center of projection, the system can be classified as noncentral projection system and central projection system. Everything has its advantages and disadvantages. In the noncentral system, the relative position and orientation between the camera and the mirror can be arbitrary which allows zooming and resolution enhancing in some selected regions of the image. However, such flexible configuration results in high complexity for system modeling. Therefore, there is no robust linear calibration algorithm for them and thus their applications often require less accuracy. The central system requires careful configuration of the camera and the mirror to ensure unique projection center. This cost brings a closed-form expression for the projection modeling, which maps 3D space points to image pixels. Hence, the complexity of system modeling has been considerably reduced. So we will focus on this kind of vision systems in this chapter, concerning with their geometrical properties and approaches of calibration and 3D reconstruction.

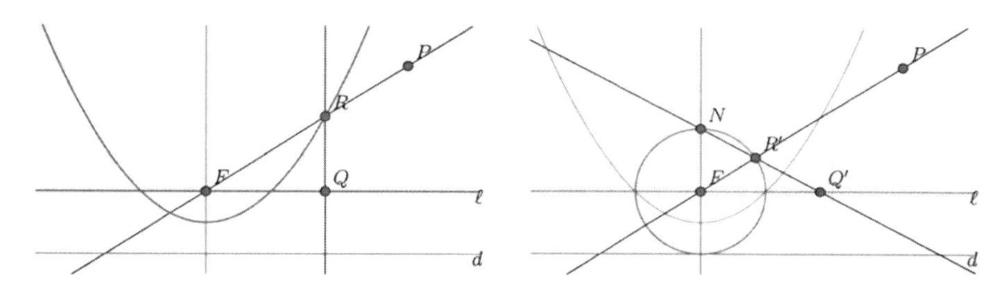

Fig. 6.5 Parablic projection is equivalent to the projection via a sphere at the north pole

6.1.2 Calibration of Omni-Directional Vision System

In the omni-directional vision system, the imaging process can be intuitively divided into two steps: firstly, 3D rays are reflected by the mirror surface following the Snell's law at the intersection points; then the reflected rays are acquired by the CCD camera into an image. In general, the first step is nonlinear mathematically since the nonlinearity in the equation of curved mirror surface. This is a critical problem for many practical uses of the systems and special efforts should be paid on it.

The geometric properties of the image formation process had been extensively studied by Baker [209] and Geyer [210] respectively. They showed that a central catadioptric projection is equivalent to a two-step mapping via the sphere: firstly, the 3D space points are projected to the sphere surface via its center; then the image is acquired via the optical center of the camera (Fig. 6.5). Some researchers, e.g. Barreto [211] and Ying [212], had proposed a unifying projection model covering the cataoptric, dioptric and fisheye cameras. For example, Barreto in [211] established a representation of the resulting image planes in the five-dimensional projective space through the Veronese mapping. In this space, a collineation of the original plane corresponds to a collineation of the lifted space.

With these geometrical theories, some calibration constraints and very practical implications can be obtained for the system. The existing techniques can be roughly divided into two classes. The first one is calibrating using image of lines [192, 193]. Mei [194] presented an algorithm for structure from motion using 3D line projections in a central catadioptric camera. With the images of three lines, Geyer [195] gave a closed-form solution for focal length, image center and aspect ratio for skewless cameras and a polynomial root solution in the presence of skew. Barreto [196] investigated the projective invariant properties of central catadioptric projection of lines and suggested an efficient method for recovering the image of the absolute conic, the relative pose between the camera and the mirror and the shape of the reflective surface. However, the estimation of the conics as imaged lines in these algorithms is hard to accomplish which limits its use in practice. Firstly, only a small part of the arc of the conic is available from the image due to occlusions and curvature, which makes the results unreliable and inaccurate. Secondly, when multiple lines are projected in the images, it is difficult to distinguish which conic is the image of a given line. Lastly, the conics are not always line images. They may be the image of lines, circles or any other curves.

The other class is to use points as important features for the calibration, which completely overcome the above listed limitations [197–199]. Most of the relevant methods consist in first calculating the epipolar geometry through fundamental matrix or homography matrix, and then extracting the involved parameters, such as mirror parameters, camera intrinsic or extrinsic parameters [200–204]. Svoboda [200] firstly proposed an adapted version of the classical 8-point algorithm for Panoramic Cameras. Micusik [202] showed that the epipolar geometry can be estimated from a small number of correspondences by solving a polynomial eigenvalue problem. Mei [203] tracked planar templates and calculated the homography matrix to compute the camera motion. Due to the large field of view and distortion of images, it is often difficult to obtain high accurate calibration of the catadioptric system. So Lhuillier [204] and Goncalves [205] recently suggested bundle adjustment to improve the accuracy and good experimental results are obtained. Since nonlinear optimization is involved, the computation cost of their algorithms is inefficient.

6.1.3 Test Example

In catadioptric system, special geometric properties are observed. For example, Geyer and Daniilidis in [192] pointed out that the intersection of two circles which are projections of parallel lines depends only on the common direction and not on the positions of the parallel lines, and the image centers of those lines are collinear. This implies that the intersections are actually the vanishing point of those lines.

Here, we have done an experiment to test this observation. Figure 6.6a gives randomly generated three parallel lines, each contains thirty 3D points. Their images in a simulated catadioptric system are shown in Fig. 6.6b. Then conic fitting is performed with all the imaged points and the results are given in Fig. 6.6c, in which the ' + ' and the solid '.' represents the centers and the imaged points, respectively. We can see that there are two intersections for the three conics and the centers are collinear as expected. Our experiments further demonstrate that they are invariant to the positions of the parallel lines. Now, Gaussian noise with zero mean and different levels of standard deviations is injected to the imaged points to test the robustness. We find that the conic fitting is very sensitive to the noisy data. Figure 6.6d illustrates one of the results with the noise level equal to 0.4. This simulation tells us that more representative points as possible should be involved when calibrating a catadioptric system with 3D lines. However, it is not easy to satisfy such requirement since a line segment only occupies a very small portion in the image. Hence, we will use points as important features in the following sections.

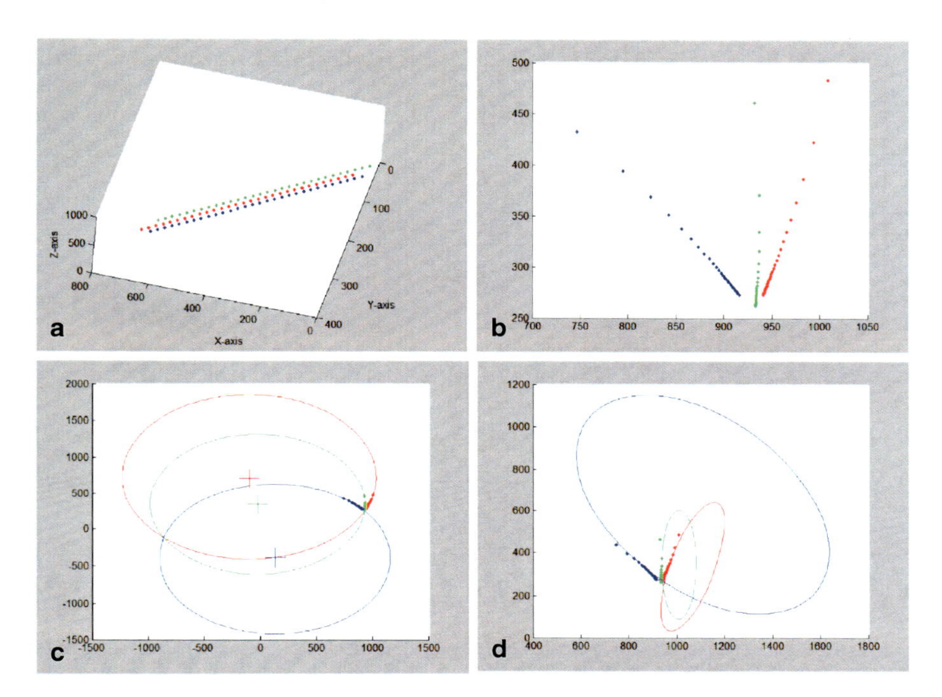

Fig. 6.6 Images of parallel lines in the simulation system. **a** Three parallel lines in 3D space. **b** Ideal images in a catadioptric system. **c** Conic fittings with the imaged points. **d** Conic fitting in case of noisy points

6.2 Panoramic Stereoscopic System

6.2.1 System Configuration

Figure 6.7a illustrates the configuration of the panoramic stereoscopic system, which consists of a fisheye lens camera and a convex reflective mirror. The optical axis of the fisheye lens camera and the optical axis of the mirror are aligned. The system views its surroundings partly through the fisheye lens (from vertical view angle α_1 to α_2) and partly through the reflection of the convex mirror (from vertical view angle β_1 to β_2). It is easy to see that the region of $\Delta\theta$ represents the overlap between the two fields of views. Within this region, since an object of interest can be observed simultaneously by the system from separated views, its 3D position can be reconstructed from a single image by stereoscopic vision method. Hence, the system is named as panoramic stereoscopic system. It should be note that the illustration in Fig. 6.7a describes the case when the catadioptric image satisfies a single viewpoint (SVP) constraint. For

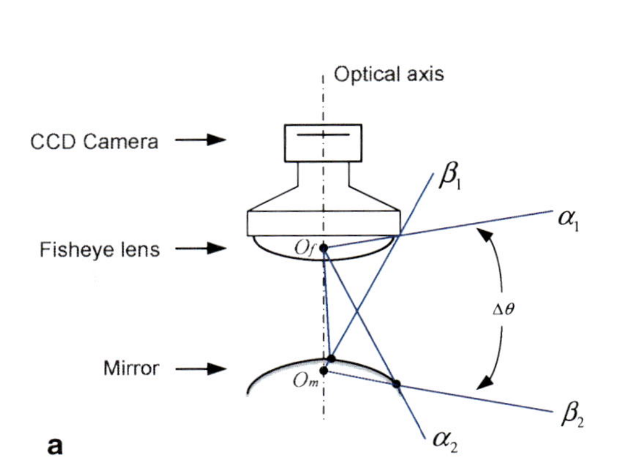

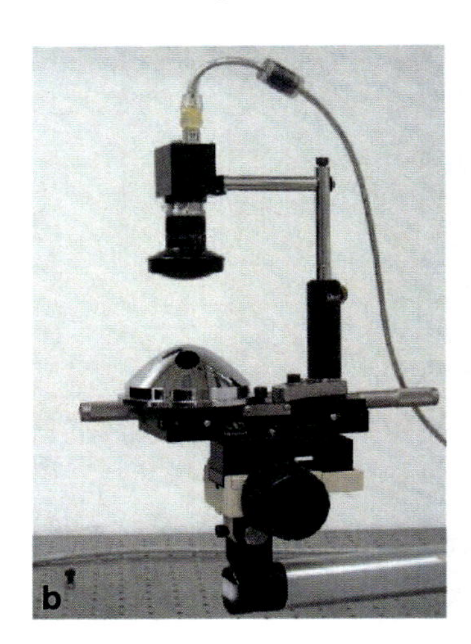

Fig. 6.7 Configuration of the panoramic stereoscopic system

non-SVP case, the viewpoint in the mirror would move along the optical axis within a range defined by a caustic [209], which would be further explained later.

A prototype of the system has been constructed as shown in Fig. 6.7b. All the components used in the system are off-the-shelf components and are connected onto an optical mounting stage. A color CCD video camera with 1394 interface is used with an image resolution of 1360 pixels by 1024 pixels. A Fujinon 1.4 mm fisheye lens with a field of view of 185° is mounted to the camera. (The actual field of view of this lens is found to be 194°.) We choose a hyperbolic mirror as the convex mirror due to two main reasons: the first is that it provides a larger vertical field of view; the second is that it is easy to construct the system with central projection since the mirror has two focal points. In our system, the mirror is provided by ACCOWLE Company as part of an omnidirectional sensor.

Figure 6.8a shows an image taken by the prototype system in the middle of a square shaped room. The image can be decomposed into four concentric ring shaped image regions as marked by A, B, C, and D in Fig. 6.8b. Region A and region B are captured through the mirror and formed the catadioptric image while region C and region D are captured through the fisheye camera and form the fisheye image. We find that region B and D have an overlapping field of view and form the stereo image pair in the image.

A latitude-longitude unwrapping of Fig. 6.8 is shown in Fig. 6.9. This image is composited by stitching the common in image region B and D. The upper part is from region A and the lower part is from region C while the middle part is from region B or D. As the vertical field of view is extended, both the ceiling and floor becomes visible. Here only region B is displayed for the overlapping view.

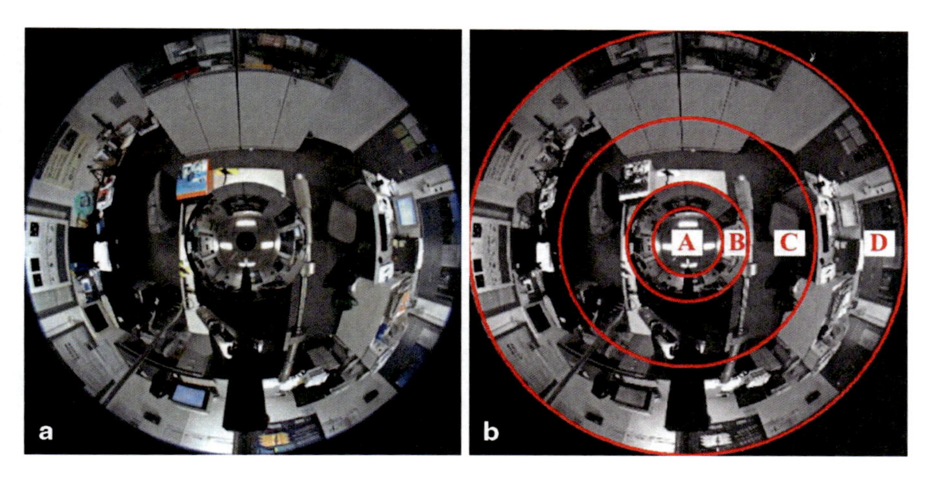

Fig. 6.8 Image of an indoor environment captured by the system

Fig. 6.9 A latitude-longitude unwrapping of the image

6.2.2 Co-axis Installation

A desired feature in the system installation is the alignment between the axis of the mirror and the optical axis of the fisheye lens. In general, this is not a trivial task as neither the two axes are directly measurable. In this work, the mirror manufacturer provides a screw hole at the mirror's optical axis with high localization precision, which can be used as a marker for the alignment task. Figure 6.10a shows a side view of this screw hole. With an ideal installation as shown in Fig. 6.7, the image of this screw hole and the image of the fisheye lens are supposed to coincide at the same center in a captured image. We make use of this observation as an indicator and manually adjust the setup until this constraint is satisfied as shown in Fig. 6.10b. The image of the screw hole appears as a small bright circle in the central part. The black circular region around it is the image of the fisheye lens reflected in the mirror. The color contrast makes it easy to observe their relative positions. Any slight misalignment would lead to a displacement of the screw hole away from the center of the fisheye lens, an example of which is shown in Fig. 6.10c.

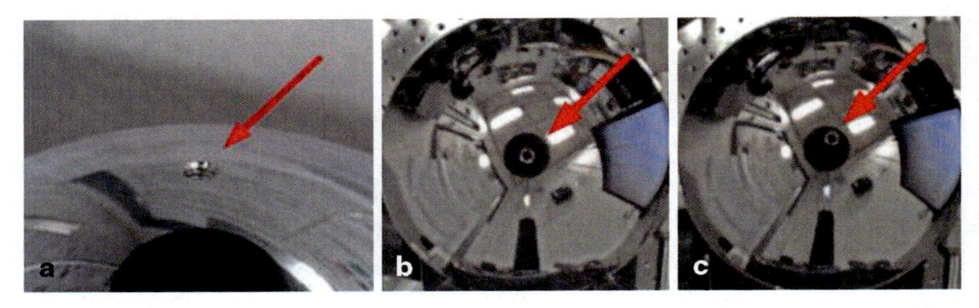

Fig. 6.10 Co-axis installation using the central screw hole as a marker. **a** A side view of the central screw hole on the mirror surface. **b** A view of the screw hole in the image when the system is co-axis installed. **c** A view of the screw hole in the image when the alignment is not strictly co-axis

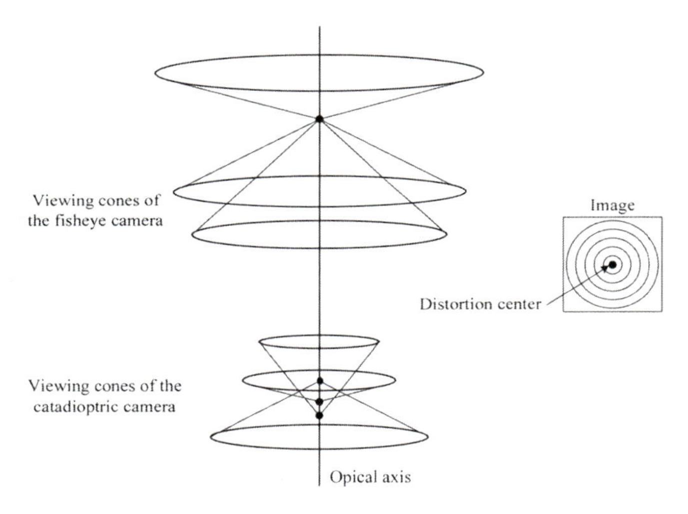

Fig. 6.11 An illustration of the generic radial distortion model

6.2.3 System Model

An illustration of the generic radial distortion model for the proposed system is shown in Fig. 6.11. The following assumptions are made for the concerned system: (1) The optical axis of the fisheye camera and the axis of the hyperbolic mirror are aligned. (2) Both the optical property of the fisheye lens and the hyperbolic mirror are rotational symmetric with respect to the optical axis. With these assumptions, the imaging geometry is radially symmetric with respect to the optical axis. Then an image distortion center can be defined as the intersection of the optical axis with the image plane. With regard to this distortion center, the image can be decomposed into a series of distortion circles. The 3D light rays associated with those pixels lying on the same distortion circle lie on a right viewing cone centered on the optical axis. Each viewing cone can be considered as an individual perspective camera with a varying focal length $f(d)$, where d is the radial distance to the distortion center. Therefore, the camera can be described by all these viewing cones.

As an image contains distortion circles from both the fisheye part and the catadioptric part, the viewing cones can be grouped into two clusters. One cluster of viewing cones describes the fisheye image and the other cluster of view cones describes the catadioptric image. This model as described above is different from the traditional pinhole models and is also different from the fully generic model [207]. It is referred as a generic radial distortion model [208], as the light ray corresponding to a pixel is only determined by the pixel's radial distance to the distortion center.

Following this, the focal length of a viewing cone $f(d)$ can be regarded as a function of d and written as:

$$f(d) = \begin{cases} f_c(d), \ d < B \\ f_f(d), \ d \geq B \end{cases} \tag{6.1}$$

Here B is the radius corresponds to the boundary between the fisheye image and the catadioptric image. Note that since both the fisheye image and catadioptric image have field of views larger than 180°, $f_c(d)$ and $f_f(d)$ may take zero values and become invertible. To fix this, the opening angle of viewing cone can be used instead. Denote half of the opening angle of a viewing cone as θ then $\tan(\theta) = d/f(d)$.

Assuming the optical property of the system is continuous, we can parameterize the angle function using polynomial assumption:

$$\theta(d) = \begin{cases} \theta_c(d) = a_0 + a_1 \cdot d + \cdots + a_n \cdot d^n, d < B \\ \theta_f(d) = b_0 + b_1 \cdot d + \cdots + b_m \cdot d^m, d \geq B \end{cases} \tag{6.2}$$

When $d = 0$, the light ray coincides with the optical axis and the opening angle of this special viewing cone becomes zero. This constraint enforces the constant component of the polynomial to be zero and thus we set $a_0 = \theta_c(0) = 0$ and $b_0 = \theta_f(0) = 0$.

In this section, we assume that the fisheye image satisfies the single-view-point (SVP) constraint. Therefore all the viewing cones of the fisheye image share the same vertex. If the catadioptric image also satisfies a SVP constraint, the distance between the optical centers of the fisheye camera and the catadioptric camera can be described by a baseline as that of a traditional stereoscopic system. However, the SVP constraint is satisfied only when the optical center of the fisheye lens is configured at the second focus of the hyperbolic mirror. Except for this configuration, non-SVP has to be considered. According to the theory presented in [209], the viewpoint variation along the optical axis can be predicted as a simulation shown in Fig. 6.12. The vertex of a viewing cone is the intersection of the optical axis with the reverse extension of a reflected light ray. It can be seen that as the radial distance d increases, the viewpoint would move outward from inside of the mirror along the optical axis.

Assume the optical center of the fisheye camera to be an origin on the optical axis. Then we model the position of a viewing cone's vertex relative to this origin as a polynomial:

$$v(d) = \begin{cases} v_c(d) = e_0 + e_1 \cdot d + \cdots + e_w \cdot d^w, & d < B \\ v_f(d) = 0, & d \geq B \end{cases} \tag{6.3}$$

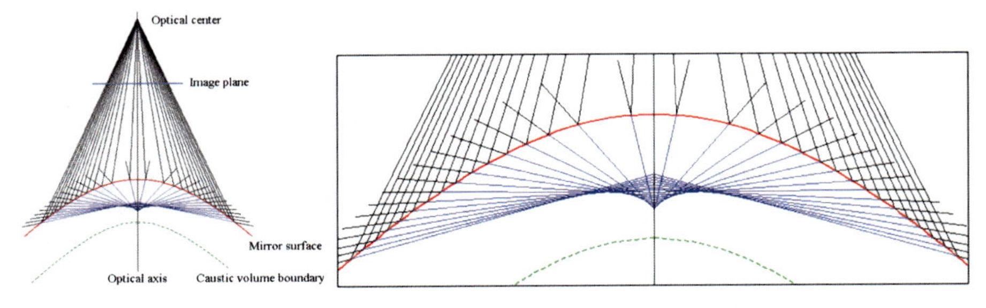

Fig. 6.12 Non-single viewpoint property of a hyperbolic mirror. *Blue lines* are the backward extension of reflected light rays

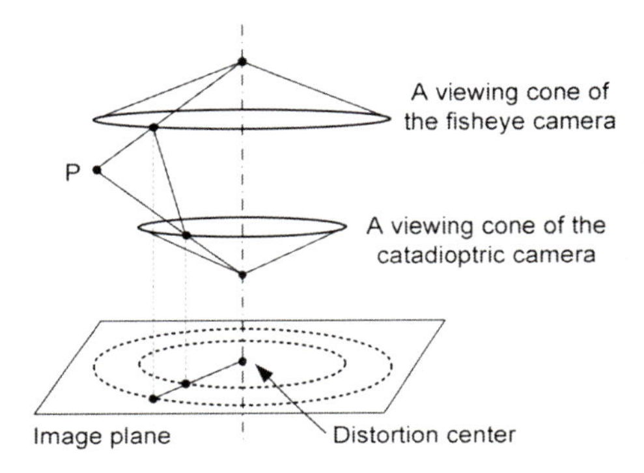

Fig. 6.13 Illustration of the epipolar geometry with the generic radial distortion model

Notice that when the high-order coefficients become zeros the function $v_c(d) = e_0$ becomes a constant, which is exactly the case of a SVP model.

In summary, the generic radial distortion model is fully expressed as the location of the distortion center, the focal length function $f(d)$ (equivalently $\theta(d)$), and the vertex position function $v(d)$ for the set of viewing cones. The number of parameters therefore depends on the ranks of the polynomial parameterization and is $2 + n + m + (w + 1)$.

6.2.4 *Epipolar Geometry and 3D Reconstruction*

With the co-axis installation, the epipolar geometry is greatly simplified to lines that pass the distortion center as shown in Fig. 6.13. Therefore, the distortion center can be determined as the intersecting point of all the line segments that connecting the image corresponding pairs. It is obvious that this geometric model is much simpler and more computationally efficient compared with the iterative method [208].

Following this epipolar geometry, a closed form solution for 3D position of an object point can be derived based on the generic radial distortion model.

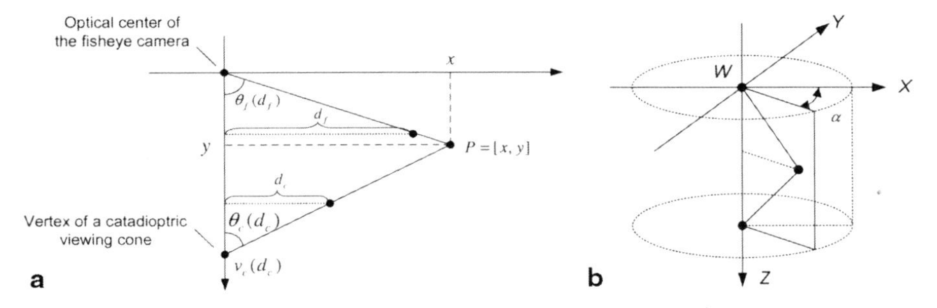

Fig. 6.14 Triangulation with the generic radial distortion model for 3D reconstruction

Let $[u_f, v_f]$ and $[u_c, v_c]$ be a pair of corresponding image points in the fisheye image and in the catadioptric image respectively. Their radial distance to the distortion center $[u_{center}, v_{center}]$ is d_c and d_f. Then the position $P = [x, y]$ of this point in a plane *x-y* determined by this point and the optical axis can be obtained by triangulation as shown in Fig. 6.14a.

Consider a world coordinate *W-XYZ* as shown in Fig. 6.14b. The Z-axis is aligned with the optical axis and the origin W is set at the optical center of the fisheye lens camera. Denote the angle between the plane *x-y* and the plane *X-Z* to be α and it can be obtained as $\alpha = \arctan((v_f)/(u_f))$. Then the 3D position of the considered point in coordinate *W-XYZ* is: $X = x\cos(\alpha)$, $Y = x\sin(\alpha)$ and $Z = y$. Following these, we give the expression of *X*, *Y*, and *Z* as:

$$\begin{aligned} X &= \frac{v(d_c)\tan(\theta_c(d_c))\tan(\theta_f(d_f))\cos(\arctan((v_f)/(u_f)))}{\tan(\theta_c(d_c)) + \tan(\theta_f(d_f))} \\ Y &= \frac{v(d_c)\tan(\theta_c(d_c))\tan(\theta_f(d_f))\sin(\arctan((v_f)/(u_f)))}{\tan(\theta_c(d_c)) + \tan(\theta_f(d_f))} \\ Z &= \frac{v(d_c)\tan(\theta_c(d_c))}{\tan(\theta_c(d_c)) + \tan(\theta_f(d_f))} \end{aligned} \tag{6.4}$$

Following the above equations, once the parameters of the camera model is calibrated and the point correspondence are known, the 3D position of a point can be explicitly computed. These equations are used to perform a non-linear optimization as introduced in the next section. Following this expression, the error of 3D reconstruction as a function of the pixel can also be explicitly expressed and analyzed.

6.2.5 Calibration Procedure

When calibrating the system, an initial estimation of the system parameters, such as the focal length and the relative poses, etc, is needed. This process is accomplished using a homography-based method [208]. As a dense correspondence is needed in

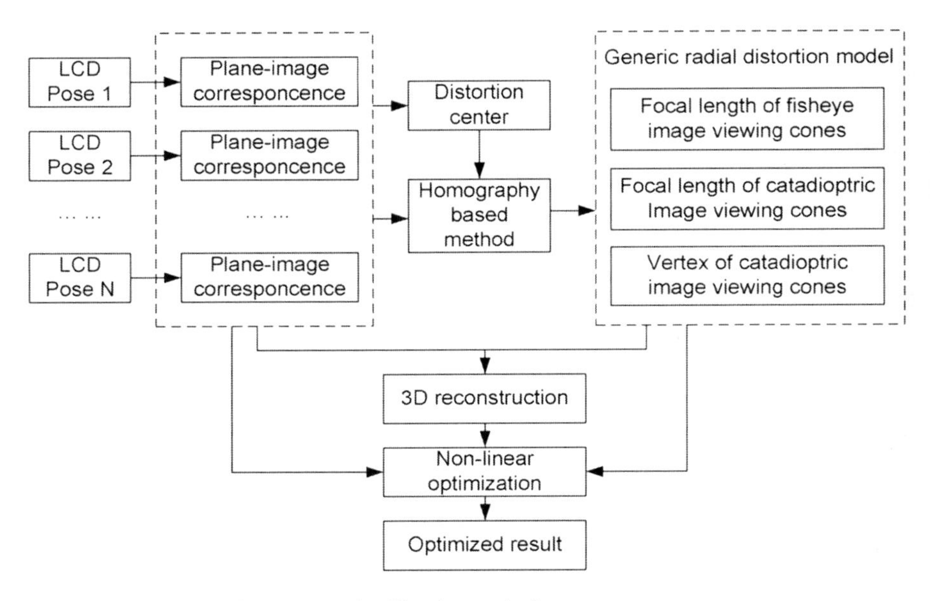

Fig. 6.15 Flow chart of the proposed calibration method

this method, it is obtained by using a LCD as an active calibration plane at several unknown poses. With this initial estimation, the 3D reconstruction capability of the system is explored and a non-linear formalization method is used to find an optimal estimation. In summary, the flow chart for the calibration procedure is given in Fig. 6.15.

6.2.5.1 Initialization of the Parameters

Here, a distortion circle of radius d in the generic radial distortion model is regarded as an independent perspective camera. At this stage, the high-order coefficients in those polynomial approximations can be regarded as zero values. With the dense correspondence, the distortion center can be directly obtained using the epipolar constraint as shown in Fig. 6.14. From the correspondence between pixels on this circle and points on the LCD plane, it is possible to compute an image-to-plane Homography H_d. Then the focal length of the considered distortion circle, rotation matrix and translation vector describing the camera pose can be obtained with this matrix. The relative pose between two viewing cones can be straightforwardly obtained since their relative poses with respect to the same calibration plane are known. This way, vertex of the catadioptric viewing cones can be calculated as translations along the optical axis. Consequently, the initial values for the parameters in the generic radial distortion model can be obtained.

6.2.5.2 Non-Linear Optimization

The model parameters given by the homography based method are improved using a non-linear optimization method in a similar way as bundle adjustment. The goal of optimization is to minimize 3D reconstruction errors for sample points on a set of N LCD planes at unknown positions. A grid of sample points are selected from each of the LCD planes and their positions are thus known as ground truth. The point positions on the kth LCD plane is denoted as the set $\{P_t{}^{(k,i)} = [X_t{}^{(k,i)}, Y_t{}^{(k,i)}, Z_t{}^{(k,i)}]\}$, $i = 1, 2, ..., M_k$, where M_k is the number of sample points on the kth plane. The local coordinates for each set $\{P_t{}^{(k,i)}\}$ selects its LCD plane as the X-Y plane and thus $Z_t{}^{(k,i)} = 0$ for each point.

As LCD can provide dense plane-image correspondence, the image points of $\{P_t{}^{(k,i)}\}$ on the image can be automatically extracted with precision and reliability. The corresponding image point of $P_t{}^{(k,i)}$ is $p_f{}^{(k,i)}$ in the fisheye image and is $p_c{}^{(k,i)}$ in the catadioptric image. The set of ground truth points $\{P_t{}^{(k,i)}\}$ and their correspondence in the image $\{p_f{}^{(k,i)}\}$ and $\{p_c{}^{(k,i)}\}$ are used as input data to the optimization algorithm.

The error function to minimize is constructed as the 3D reconstruction errors of the points on the N calibration planes as follows:

$$\begin{aligned} E = \sum_{k=1}^{N} \sum_{i=1}^{M_k} \Big(& \left(X(p_f{}^{(k,i)}, p_c{}^{(k,i)}) - T_X{}^{(k)}(X_t{}^{(k,i)})\right)^2 \\ & + \ldots \left(Y\left(p_f{}^{(k,i)}, p_c{}^{(k,i)}\right) - T_Y{}^{(k)}\left(Y_t{}^{(k,i)}\right)\right)^2 \\ & + \ldots \left(Z\left(p_f{}^{(k,i)}, p_c{}^{(k,i)}\right) - T_Z{}^{(k)}\left(Z_t{}^{(k,i)}\right)\right)^2 \Big) \end{aligned} \tag{6.5}$$

Here $X(p_f{}^{(k,i)}, p_c{}^{(k,i)})$, $Y(p_f{}^{(k,i)}, p_c{}^{(k,i)})$, and $Y(p_f{}^{(k,i)}, p_c{}^{(k,i)})$ are the 3D reconstruction using the formula 4, where the model parameter θ_f, θ_c, and v_f are involved in the computation. A rigid transformation $T^{(k)}$ transform the coordinates of a ground truth 3D point $P_t{}^{(k,i)}$ from local coordinate on the kth LCD plane to the system's camera coordinate by $T^{(k)}(P_t{}^{(k,i)}) = \mathbf{R}^{(k)} P_t{}^{(k,i)} + \mathbf{t}^{(k)}$, where $\mathbf{R}^{(k)}$ and $\mathbf{t}^{(k)}$ are the rotation matrix and a translation vector that describe the motion between the kth LCD and the camera's coordinate frame. Here $\mathbf{R}^{(k)}$ is mapped to a triple $(r_1{}^{(k)}, r_2{}^{(k)}, r_3{}^{(k)})$ using Rodrigues formula to ensure the constraints of a rotation matrix. Using the initial values of model parameters, 3D reconstruction can be performed. Then the initial values of $\mathbf{R}^{(k)}$ and $\mathbf{t}^{(k)}$ can be obtained by using a least square method that best align the reconstructed 3D data and the ground truth points for each LCD. The variables in the error function are summarized in Table 6.1.

The error function (6.5) formulized above consists of a sum of square terms. The number of square terms is $N \times M_k \times 3$ and the number of variables is $m+n+w+6 \times N$. Minimizing this error function is a non-linear minimization problem, which can be solved using the Levenberg-Marquardt algorithm. Given the initial values for the variables as listed in Table 6.1, an optimal solution can be obtained. Interested readers can refer to our recent work in [213] for some interesting experimental results.

Table 6.1 A summary of the variables in the non-linear optimization

	Expression	Variables
θ for fisheye VC[a]	$\theta_c(d) = a_0 + a_1 \cdot d + \cdots + a_n \cdot d^n$	$a_1, a_2, \ldots, a_n$
θ for catadioptric VC	$\theta_f(d) = b_0 + b_1 \cdot d + \cdots + b_m \cdot d^m$	$b_1, b_2, \ldots, b_m$
Vertex for catadioptric VC	$v_c(d) = e_0 + e_1 \cdot d + \cdots + e_n \cdot d^n$	$e_0, e_1, \ldots, e_w$
Pose of LCD 1	$T^{(1)}(P_t^{(1,i)}) = \mathbf{R}^{(1)} P_t^{(1,i)} + \mathbf{t}^{(1)}$	$(r_1^{(1)}, r_2^{(1)}, r_3^{(1)})$, $(t_1^{(1)}, t_2^{(1)}, t_3^{(1)})$
Pose of LCD 2	$T^{(2)}(P_t^{(2,i)}) = \mathbf{R}^{(2)} P_t^{(2,i)} + \mathbf{t}^{(2)}$	$(r_1^{(1)}, r_2^{(1)}, r_3^{(1)})$, $(t_1^{(1)}, t_2^{(1)}, t_3^{(1)})$
...	...	...
Pose of LCD N	$T^{(N)}(P_t^{(N,i)}) = \mathbf{R}^{(N)} P_t^{(N,i)} + \mathbf{t}(N)$	$(r_1^{(N)}, r_2^{(N)}, r_3^{(N)})$, $(t_1^{(N)}, t_2^{(N)}, t_3^{(N)})$

[a]VC is an abbreviation for viewing cone

6.3 Parabolic Camera System

In Sect. 6.2, the system is basically modeled by polynomial approximation with over-parameterization, followed by nonlinear optimization process. In practice, this is a difficult task to solve, requiring expensive computation cost and maybe trapped into local minima in some cases. So in this section, we will talk about another catadioptric system, i.e. parabolic camera system, in which exact theoretical projection function can be established.

6.3.1 System Configuration

The configuration of the parabolic camera system is illustrated in Fig. 6.16. The system consists of a paraboloid mirror and a CCD camera with orthographic projection. The mirror is also provided by ACCOWLE Company. According to the Snell's law, all the 3D rays reflected by the mirror surface will be parallel to each other, then the reflected rays are sensed by the camera into an image as shown in Fig. 6.16a. We place a telecentric-like lens between the camera and the mirror to approximate the orthographic projection.

Co-axis Installation Similar to the panoramic stereoscopic system, the optical axis of the camera should be aligned with the symmetric axis of the mirror. This is simply ensured in this system since those off-the-shelf components including the CCD, the telecentric lens and the mirror, are tightened by screws and they meet the symmetry requirement by manufacturing. For example, the pin on the mirror surface provided by the manufacturer can be act as the symmetry axis of the mirror with high precision, as shown in Fig. 6.16b.

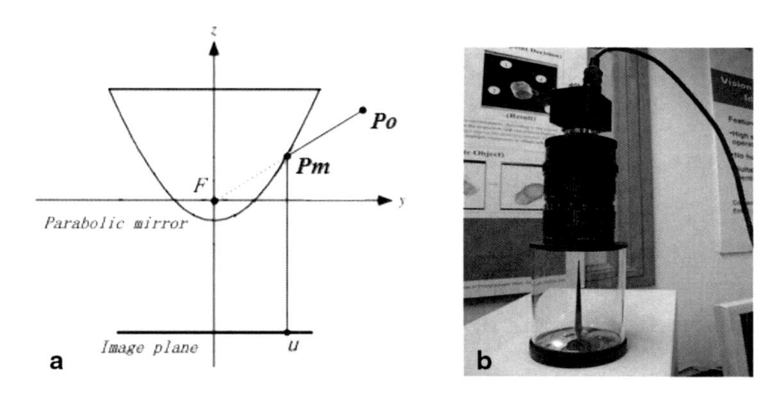

Fig. 6.16 Parabolic camera system. **a** Geometry in the system. **b** A prototype of the parabolic camera system

6.3.2 System Modeling

Three right-handed coordinate systems are firstly established, i.e. the world, the mirror and the camera coordinate systems. The world coordinate system is attached the concerned scene or objects. The origins of the mirror and the camera coordinate systems coincide with the focal point of the mirror and the optical center of the camera, respectively. The transformation between them is described as rotation matrix $\boldsymbol{R}_c$ and $\boldsymbol{t}_c$ while the transformation between the world and the mirror coordinate systems is denoted by $\boldsymbol{R}_m$ and $\boldsymbol{t}_m$.

With those coordinate systems, the image formation can be divided into two steps: (1) The world point, say $\boldsymbol{P}_0$, is projected onto the mirror surface $\boldsymbol{P}_m$ by a central projection from the focal point $\boldsymbol{F}$; (2) Then point $\boldsymbol{P}_m$ is projected into the image plane of the camera from the optical center to obtain a pixel u. In this model, all the parallel reflected rays can be considered as passing through the point at infinity. Hence, the system is deemed as central catadioptric system that owns a single effective viewpoint. We assume that the intrinsic parameters of camera have been calibrated using the algorithm suggested in Chap. 3. For clarity, we suppose that the system is stationary while the concerned object performs a rigid motion. The task in this section becomes how to recover the mirror parameter and the motion parameters.

For simplicity, we assume that the optical axis of the camera is aligned with the symmetric axis of the mirror. So we can set $\boldsymbol{R}_c = \boldsymbol{I}$ and $\boldsymbol{t}_c = 0$.

Let the surface equation of the parabolic mirror be

$$z = \frac{x^2 + y^2 - a^2}{2a} \tag{6.6}$$

where a is called the mirror parameter.

Suppose an arbitrary point in the object and its corresponding point on the mirror surface are respectively denoted by $\boldsymbol{P}_o$ and $\boldsymbol{P}_m$. It is obvious that their relationship can be described by

$$\boldsymbol{P}_m = \lambda \boldsymbol{P}_o \tag{6.7}$$

where λ is a scale factor.

Let the coordinates of the 3D point $\boldsymbol{P}_o$ be $\boldsymbol{P}_o = [x\ y\ z]^{\mathrm{T}}$. Substituting (6.7) into (6.6), we have two solutions for λ, corresponding the two intersections between the incident ray and the mirror surface. As shown in Toepfer [206], the value of λ should be positive, i.e. $\lambda = \frac{a(z+\sqrt{x^2+y^2+z^2})}{x^2+y^2}$. Hence, the point $\boldsymbol{P}_m$ on mirror can be determined.

Then the orthographic projection from point $\boldsymbol{P}_m$ to the image point $\boldsymbol{u}$ can be written as

$$\boldsymbol{u} = \begin{bmatrix} p_1 \\ p_2 \\ 1 \end{bmatrix} \tag{6.8}$$

where $\boldsymbol{p}_i$ represents i-th element in $\boldsymbol{P}_m$.

Now, let the object perform an arbitrary rigid motion. So the point $\boldsymbol{P}_o$ is moved to the position of $\boldsymbol{p}'_o$. Similarly, we can have

$$\boldsymbol{P}'_m = \lambda' \boldsymbol{P}'_o \tag{6.9}$$

Then the image pixel $\boldsymbol{u}'$ can be obtained following the same way in (6.7) and (6.8).

On the other hand, if an image point $\boldsymbol{u}$ is known, the corresponding point on the mirror surface can be directly computed as

$$\boldsymbol{P}_m = \begin{bmatrix} u \\ v \\ \frac{r-a^2}{2a} \end{bmatrix} \tag{6.10}$$

where u and v denote the two components of $\boldsymbol{u}$, $r = u^2 + v^2$ for conciseness.

In fact, (6.10) shows the back-projection process transferring from an image point to the mirror point, and hence the related 3D light ray, i.e. $\boldsymbol{F}\boldsymbol{P}_m$. We can then obtain the 3D space point by traditional triangulation method if two or more such rays are estimated. This is generally named as the 3D reconstruction.

6.3.3 Calibration with Lifted-Fundamental-Matrix

In the catadioptric system, the high nonlinearity in the imaging process is an important aspect and should be addressed properly. We will show here that the nonlinearity can be encoded into a 4×4 lifted fundamental matrix.

6.3.3.1 The Lifted Fundamental Matrix

Defining a 3 × 4 matrix $\boldsymbol{T} = \begin{bmatrix} 1 & 0 & 0 & 0 \\ 0 & 1 & 0 & 0 \\ 0 & 0 & \frac{1}{2a} & -\frac{a}{2} \end{bmatrix}$ and $\widetilde{\boldsymbol{u}} = \begin{bmatrix} u \\ v \\ r \\ 1 \end{bmatrix}$.

From (6.10), we can have

$$\boldsymbol{P}_m = \boldsymbol{T}\widetilde{\boldsymbol{u}} \tag{6.11}$$

Similarly, we obtain

$$\boldsymbol{P}'_m = \boldsymbol{T}\widetilde{\boldsymbol{u}}' \tag{6.12}$$

Here, the matrix $\boldsymbol{T}$ is called the transition matrix since it transits a pixel in the image to the corresponding point on the mirror surface.

Let the motion of the object be denoted by rotation matrix $\boldsymbol{R}$ and translation vector $\boldsymbol{t}$. Then we have

$$\boldsymbol{P}'_o = \boldsymbol{R}\boldsymbol{P}_o + \boldsymbol{t} \tag{6.13}$$

Let the skew-symmetric matrix of vector $\boldsymbol{t}$ be $[\boldsymbol{t}]_\times = \begin{bmatrix} 0 & -t_3 & t_2 \\ t_3 & 0 & -t_1 \\ -t_2 & t_1 & 0 \end{bmatrix}$.

Multiplying it on both sides of (6.13), we obtain

$$[\boldsymbol{t}]_\times \boldsymbol{P}'_o = [\boldsymbol{t}]_\times \boldsymbol{R}\boldsymbol{P}_o \tag{6.14}$$

It is well known that $\boldsymbol{v}^{\mathrm{T}}[\boldsymbol{t}]_\times \boldsymbol{v} = 0$ for any 3 × 1 vector v.

So left multiplying by $\boldsymbol{P}_o^{'\mathrm{T}}$ on both sides of (6.14), we get

$$\boldsymbol{P}_o^{'\mathrm{T}} \boldsymbol{E}\boldsymbol{P}_o = 0 \tag{6.15}$$

where $\boldsymbol{E} = [\boldsymbol{t}]_\times \boldsymbol{R}$ is a 3 × 3 matrix. It contains the information of rotation matrix and translation vector.

It is obvious that $\boldsymbol{E}$ is something like the essential matrix in traditional stereo vision. But we cannot compute it directly since we yet do not know the point pairs ($\boldsymbol{P}_o$ and $\boldsymbol{P}'_o$).

Considering (6.7), (6.9), (6.11), (6.12) and (6.15), we have

$$\widetilde{\boldsymbol{u}}^{'\mathrm{T}} \widetilde{\boldsymbol{F}} \widetilde{\boldsymbol{u}} = 0 \tag{6.16}$$

where $\widetilde{\boldsymbol{F}} = \boldsymbol{T}^{\mathrm{T}}\boldsymbol{E}\boldsymbol{T}$ is a 4 × 4 matrix and the rank is 2. We term it the lifted fundamental matrix. It is related with the mirror shape and the motion parameters, and hence completely encodes the geometry of the parabolic camera system.

6.3.3.2 Calibration Procedure

From (6.16), the lifted fundamental matrix is a projectivity between the two images and hence can be estimated from their feature correspondences. Next, we will talk about the way for estimating the matrix.

Without loss of generality, let $\widetilde{\boldsymbol{F}} = \begin{bmatrix} f_1 & f_2 & f_3 & f_4 \\ f_5 & f_6 & f_7 & f_8 \\ f_9 & f_{10} & f_{11} & f_{12} \\ f_{13} & f_{14} & f_{15} & f_{16} \end{bmatrix}$ and $\boldsymbol{f} = [f_1\ f_2\ f_3\ \ldots\ f_{16}]^{\mathrm{T}}$ be its row-first vector.

From (6.16), each pair of corresponding pixels provides one linear constraint:

$$[uu'\ uv'\ ur'\ u\ vu'\ vv'\ vr'\ v\ ru'\ rv'\ rr'\ r\ u'\ v'\ r'\ 1]\boldsymbol{f} = 0 \tag{6.17}$$

So given n ($n \geq 16$) pairs of pixels, we have the following linear equation system:

$$\boldsymbol{A}\boldsymbol{f} = 0 \tag{6.18}$$

where $\boldsymbol{A} = \begin{bmatrix} u_1u'_1 & u_1v'_1 & u_1r'_1 & u_1 & v_1u'_1 & v_1v'_1 & v_1r'_1 & v_1 & r_1u'_1 & r_1v'_1 & r_1v'_1 & r_1r'_1 & r_1 & u'_1 & v'_1 & r'_1 & 1 \\ \vdots & \vdots & \vdots & \vdots & \vdots & \vdots & \vdots & \vdots & \vdots & \vdots & \vdots & \vdots & \vdots & \vdots & \vdots & \vdots & \vdots \\ u_nu'_n & u_nv'_n & u_nr'_n & u_n & v_nu'_n & v_nv'_n & v_nr'_n & v_n & r_nu'_n & r_nv'_n & r_nv'_n & r_nr'_n & r_n & u'_n & v'_n & r'_n & 1 \end{bmatrix}$

represents the stacked coefficient matrix.

Then the solution for the vector $\boldsymbol{f}$ can be determined up to a scale factor by SVD. Consequently, the lifted fundamental matrix can be linearly estimated from sixteen or more pairs of image pixels. This result can be optimized by the following nonlinear least squares minimization:

$$\min_{\widetilde{\boldsymbol{F}}} \sum_n \left(\frac{1}{(\widetilde{\boldsymbol{F}}\widetilde{\boldsymbol{u}}_n)_1^2 + (\widetilde{\boldsymbol{F}}\widetilde{\boldsymbol{u}}_n)_2^2} + \frac{1}{(\widetilde{\boldsymbol{F}}\widetilde{\boldsymbol{u}}'_n)_1^2 + (\widetilde{\boldsymbol{F}}\widetilde{\boldsymbol{u}}'_n)_2^2} \right) (\widetilde{\boldsymbol{u}}'^{\mathrm{T}}\widetilde{\boldsymbol{F}}\widetilde{\boldsymbol{u}})^2 \tag{6.19}$$

The cost function in (6.19) is to geometrically minimize the distances between image points and their epipolar curves, and the number of variables is sixteen. Once the lifted fundamental matrix is obtained, the mirror parameter and the motion parameters can be estimated from it. Let ε_i be the i-th element of matrix $\boldsymbol{E}$ in (6.15). Expanding the formula of $\widetilde{\boldsymbol{F}}$, we have

$$\widetilde{\boldsymbol{F}} = \begin{bmatrix} \varepsilon_1 & \varepsilon_2 & \frac{\varepsilon_3}{2a} & -\frac{a\varepsilon_3}{2} \\ \varepsilon_4 & \varepsilon_5 & \frac{\varepsilon_6}{2a} & -\frac{a\varepsilon_6}{2} \\ \frac{\varepsilon_7}{2a} & \frac{\varepsilon_8}{2a} & \frac{\varepsilon_9}{4a^2} & -\frac{\varepsilon_9}{4} \\ -\frac{a\varepsilon_7}{2} & -\frac{a\varepsilon_8}{2} & -\frac{\varepsilon_9}{4} & \frac{a^2\varepsilon_9}{4} \end{bmatrix} \tag{6.20}$$

From (6.20), we can derive at least four constraints on the mirror parameter, such as

$$a = \sqrt{-\frac{\widetilde{\boldsymbol{F}}_{41}}{\widetilde{\boldsymbol{F}}_{31}}}, \quad a = \sqrt{-\frac{\widetilde{\boldsymbol{F}}_{42}}{\widetilde{\boldsymbol{F}}_{32}}}, \quad a = \sqrt{-\frac{\widetilde{\boldsymbol{F}}_{43}}{\widetilde{\boldsymbol{F}}_{33}}}, \quad a = \sqrt{-\frac{\widetilde{\boldsymbol{F}}_{44}}{\widetilde{\boldsymbol{F}}_{34}}} \tag{6.21}$$

So we can compute the mirror parameter from any one of the above constraints or obtain the estimation in a least-squares sense with them all.

Then from (6.20), we can extract all the elements for the matrix $\boldsymbol{E}$. Consequently, we have many optional algorithms, for example [106], to decompose it into the motion parameters, i.e. rotation matrix $\boldsymbol{R}$ and translation vector $\boldsymbol{t}$.

Now, we can summarize the algorithm for calibrating the parabolic camera system into the following five steps:

Step 1: Extracting sixteen or more pairs of corresponding pixels from images and stacking them into the coefficient matrix $\boldsymbol{A}$;
Step 2: Computing the lifted fundamental matrix according to (6.18) and then optimize it using (6.19);
Step 3: Estimating the value for mirror parameter by (6.21);
Step 4: Calculating matrix $\boldsymbol{E}$ and decomposing it into the motion parameters.
Step 5: The results can be refined by bundle adjustment, after having obtained all the variable parameters.

6.3.3.3 Simplified Case

In many practical applications, planar motion or pure translation of the parabolic camera system is applicable due to its large field of view. Such special motion case with less degree of freedoms should bring considerably simpler calibration algorithm. In this section, we will discuss pure translation case which is a simulation of animal and our human vision systems.

I. Simplified Lifted Fundamental Matrix We consider the case of pure translation in the XY-plane of the world coordinate system, which means $\boldsymbol{R} = \boldsymbol{I}$ is an identity matrix and $\boldsymbol{t} = [t_1 \quad t_2 \quad 0]^{\mathrm{T}}$.

Similarly, expanding the matrix $\widetilde{\boldsymbol{F}}$ in (6.16), we have

$$\widetilde{\boldsymbol{F}} = \begin{bmatrix} 0 & 0 & \frac{t_2}{2a} & -\frac{at_2}{2} \\ 0 & 0 & -\frac{t_1}{2a} & \frac{at_1}{2} \\ -\frac{t_2}{2a} & \frac{t_1}{2a} & 0 & 0 \\ \frac{at_2}{2} & -\frac{at_1}{2} & 0 & 0 \end{bmatrix} \tag{6.22}$$

The above matrix is a skew symmetric matrix with three degree of freedoms. It is obvious that the expression is much simpler than that in (6.20). Hence, we call it the simplified lifted fundamental matrix here. Once this matrix is obtained, we can calculate the mirror parameter and the translation vector.

II. Estimation Let the simplified lifted fundamental matrix be represented by a 4-vector, i.e.

$$\boldsymbol{f} = [f_1 \quad f_2 \quad f_3 \quad f_4]^{\mathrm{T}} \tag{6.23}$$

So each pair of corresponding pixels provides one linear constraint

$$[u'r - ur' \quad u' - u \quad v'r - vr' \quad v' - v]\boldsymbol{f} = 0 \tag{6.24}$$

Given n pairs of pixels, we obtain the stacked linear equation system:

$$\boldsymbol{B}\boldsymbol{f} = 0 \tag{6.25}$$

The solution for the vector can be determined up to a scale factor by SVD or eigenvalue decomposition. Then the result can be optimized in a least-squares sense from more pairs of image pixels.

III. Calibration Procedure Once the simplified lifted fundamental matrix has been obtained, we can implement the calibration process as what follow.

From (6.2), we have two constraints on the mirror parameter

$$a = \sqrt{-\frac{\widetilde{\boldsymbol{F}}_{14}}{\widetilde{\boldsymbol{F}}_{13}}}, \quad a = \sqrt{-\frac{\widetilde{\boldsymbol{F}}_{24}}{\widetilde{\boldsymbol{F}}_{23}}} \tag{6.26}$$

Hence the mirror parameter can be estimated by any one of the above two constraints.

Then the two elements of the translation vector can be computed as

$$t_1 = \frac{2\widetilde{\boldsymbol{F}}_{24}}{a}, \quad t_2 = 2a\widetilde{\boldsymbol{F}}_{13} \tag{6.27}$$

Summarily, we can see that a similar calibration process as suggested in the previous section can be adopted. The only difference lies in that less point correspondences are required and the procedure is a bit simple in this case. The details are omitted here for conciseness.

6.3.3.4 Discussion

Compared with the panoramic stereoscopic system, the imaging process in the parabolic camera system is simple and can be modeled analytically without using polynomial approximation. This characteristic decreases the computation complexity. Both the general case and the simplified case can provide closed-form solution for the mirror parameter and the motion parameters. As it is known, the key point is to estimate the lifted fundamental matrix using corresponding points in the images of 3D scene. If those points are planar or nearly coplanar, the estimation will be degenerate. In this case, the homographic matrix can be employed, as will be discussed in the next section.

6.3.4 Calibration Based on Homographic Matrix

This section deals with the possible case when using coplanar points for calibrating the parabolic camera system. The mirror parameter and the relative position between the world frame and the mirror frame are concerned. The homographic matrix between the plane and the mirror surface is employed for this task.

6.3.4.1 Plane-to-Mirror Homography

Assuming that the world frame is attached to the 3D space plane, we consider the transformation between space plane and the mirror surface, i.e. the rotation matrix $\boldsymbol{R}_m$ and the translation vector $\boldsymbol{t}_m$.

Given a point $\boldsymbol{P}_o$ on the plane, its corresponding point on the mirror surface is:

$$\boldsymbol{P}_m = \lambda(\boldsymbol{R}_m \boldsymbol{P}_o + \boldsymbol{t}_m) \tag{6.28}$$

where λ is a nonzero scale factor.

Without loss of generality, the X-axis and Y-axis of the world frame are set to lie in the plane. So the third coordinate of $\boldsymbol{P}_o$ should be zero. Then we have the following simplified form:

$$\boldsymbol{P}_m = \lambda \boldsymbol{H} \boldsymbol{P}_o \tag{6.29}$$

where $\boldsymbol{H} = [\boldsymbol{r}_1 \ \boldsymbol{r}_2 \ \boldsymbol{t}_m]$ is named as the plane to mirror surface Homography, in which $\boldsymbol{r}_1$ and $\boldsymbol{r}_2$ are the first two columns of $\boldsymbol{R}_m$.

To estimate this Homography, the coordinates of $\boldsymbol{P}_o$ in the plane can be found from the metric information, but not for the mirror point $\boldsymbol{P}_m$. However, we can derive its expression from the corresponding image point $\boldsymbol{u}$ as in (6.10), from which partial constraints on the Homography can be firstly obtained.

6.3.4.2 Calibration Procedure

Let $\boldsymbol{P}_o = [x \ y \ 1]^{\mathrm{T}}$ and its image point $\boldsymbol{u} = [u \ v \ 1]^{\mathrm{T}}$.

We have

$$\begin{bmatrix} u \\ v \\ \frac{u^2+v^2-a^2}{2a} \end{bmatrix} = \lambda \boldsymbol{H} \begin{bmatrix} x \\ y \\ 1 \end{bmatrix} \tag{6.30}$$

In (6.30), the unknowns include the mirror parameter a, the scale factor λ and the Homographic matrix $\boldsymbol{H}$. Since it has a special form, we can partially recover the matrix $\boldsymbol{H}$ using some point pairs.

Let $\boldsymbol{H} = \begin{bmatrix} h_1 & h_2 & h_3 \\ h_4 & h_5 & h_6 \\ h_7 & h_8 & h_9 \end{bmatrix}$. From the first two rows in Eq. (6.30), we obtain

$$(h_1 x + h_2 y + h_3)v - (h_4 x + h_5 y + h_6)u = 0 \tag{6.31}$$

We can see that each point pair provides one constraint on the six components of the Homographic matrix. Given n-pair of points ($n \geq 6$), we have

$$\boldsymbol{Ah} = \boldsymbol{0} \tag{6.32}$$

where

$$\boldsymbol{A} = \begin{bmatrix} v_1 x_1 & v_1 y_1 & v_1 & -u_1 x_1 & -u_1 x_1 & -u_1 \\ \cdots & \cdots & \cdots & \cdots & \cdots & \cdots \\ v_n x_n & v_n y_n & v_n & -u_n x_n & -u_n y_n & -u_n \end{bmatrix} \quad \text{and}$$

$$\boldsymbol{h} = [h_1 \quad h_2 \quad h_3 \quad h_4 \quad h_5 \quad h_6]^{\mathrm{T}}.$$

Using eigenvalue decomposition, the solution for the vector $\boldsymbol{h}$ can be determined up to a scale factor by the eigenvector corresponding to the smallest eigenvalue of $\boldsymbol{A}$. And the scale factor s should be determined before we go on.

Considering $\boldsymbol{R}_m$ is an orthogonal matrix, we get

$$s = \frac{1}{norm\left(\begin{bmatrix} h_1 & h_2 \\ h_4 & h_5 \end{bmatrix}\right)} \tag{6.33}$$

Let R_{ij} represents the ij-th element in $\boldsymbol{R}_m$ and t_i the i-th element in $\boldsymbol{t}_m$. The first 2×2 sub-matrix of $\boldsymbol{R}_m$ can be determined as follows:

$$R_{11} = sh_1, \quad R_{12} = sh_2, \quad R_{21} = sh_4 \quad \text{and} \quad R_{22} = sh_5 \tag{6.34}$$

Consequently, the rest parameters in $\boldsymbol{R}_m$, i.e. R_{13}, R_{23}, R_{31}, R_{32} and hence R_{33}, can be estimated according to the property of rotation matrix.

For the translation vector, its first two components are computed as:

$$t_1 = sh_3, \quad t_2 = sh_6 \tag{6.35}$$

Up to now, we have recovered the rotation matrix and the partially translation vector using the Homographic matrix. In the next subsection, the rest variable parameters are estimated using the constraints from (6.30).

From the first and third rows in (6.30), we have

$$(h_1 x + h_2 y + h_3)\frac{u^2 + v^2 - a^2}{2a} - (h_7 x + h_8 y + h_9)u = 0 \tag{6.36}$$

Rearranging (6.36), we obtain

$$k^1 a^2 + uah_9 + k^2 a + k^3 = 0 \tag{6.37}$$

where $k^1 = h_1 x + h_2 y + h_3$, $k^2 = 2u(h_7 x + h_8 y)$ and $k^3 = -(u^2 + v^2)(h_1 x + h_2 y + h_3)$.

In the above equation, the unknowns include h_9 and a since h_7 and h_8 can be determined as R_{31}/S and R_{32}/S. So they can be computed with analytic solution using two pairs of point. Given $n > 2$ pairs, we can have the following linear system

$$\boldsymbol{Bb} = \boldsymbol{0} \tag{6.38}$$

where $\boldsymbol{B} = \begin{bmatrix} k_1^1 & u_1 & k_1^2 & k_1^3 \\ \cdots & \cdots & \cdots & \cdots \\ k_n^1 & u_n & k_n^2 & k_n^3 \end{bmatrix}$ and $\boldsymbol{b} = [a^2 \; ah_9 \; a \; 1]^{\mathrm{T}}$.

Proposition *The rank of matrix* $\boldsymbol{B}$ *in* (38) *is* 3.

Hence, the solution for the vector $\boldsymbol{b}$ can be obtained up to a scale factor by SVD. Then we have

$$a = \frac{b_1}{b_3} \quad \text{and} \quad h_9 = \frac{b_2}{b_3} \tag{6.39}$$

where b_i denotes the i-th element of $\boldsymbol{b}$. From the above analysis, the third component of the translation vector, i.e. t_3, is equivalent to sh_9.

Up to now, all the parameters in the parabolic camera system have been recovered analytically and the calibration task is complete. Summarily, the calibration procedure consists of the following six steps:

1. Place a planar pattern in the scene. The position is not critical since the system has a very large field of view.
2. Establish six or more point correspondences between the camera image and the planar pattern.

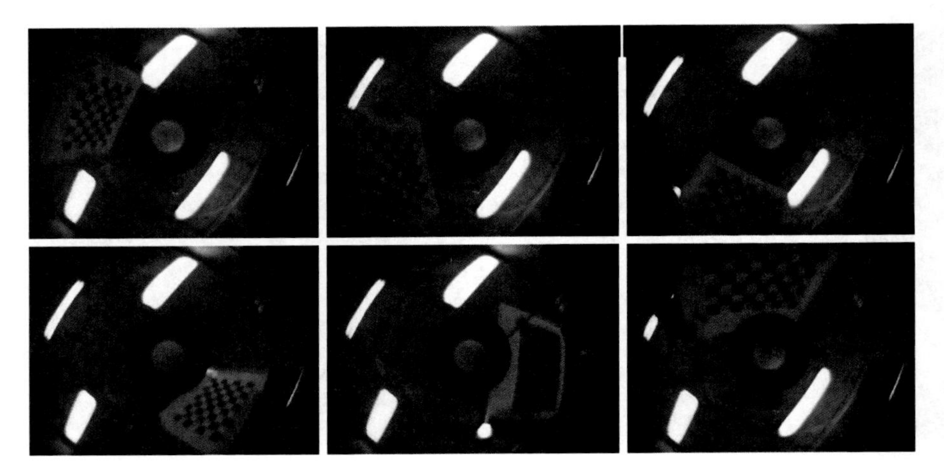

Fig. 6.17 The pattern in 6 different positions: they are labeled by 1, 2,. . . , 6 from left to right and top to down in this experiment

3. Construct the coefficient matrix A in (8) and solve it by SVD for the vector h.
4. Recover the motion parameters partially according to (10)–(12).
5. Construct the coefficient matrix B in (15) and solve it for the mirror parameters and the last component of the translation vector.
6. Optionally, the solutions are optimized by bundle adjustment using all the point correspondences.

6.3.4.3 Calibration Test

In this calibration test, the parabolic camera system given in Fig. 6.16b is employed to take the needed images. Here, a planar pattern printed in a general A4 paper is used to provide planar information. Instead of moving the vision system, we arbitrarily place the pattern around the system in our lab to capture image sequence. Figure 6.17 shows those images of the pattern in six different positions. The calibration procedure given in the previous section is performed to obtain the mirror parameter and motions of the pattern. With all the obtained parameters in the system, we can implement 3D reconstruction to visually demonstrate those results. Figure 6.18 illustrates the reconstructed positions of the pattern in the six positions. Besides, the corner points of the pattern are back-projected into the images and the distances between the back-projected points and the corresponding image points in the 2D image space are calculated. Table 6.2 gives the means and standard deviations of those distances.

6.3.5 Polynomial Eigenvalue Problem

In the previous section, metric information from the scene plane is needed to estimate the homographic matrix. So it is inapplicable for an arbitrary scene with planar patch.

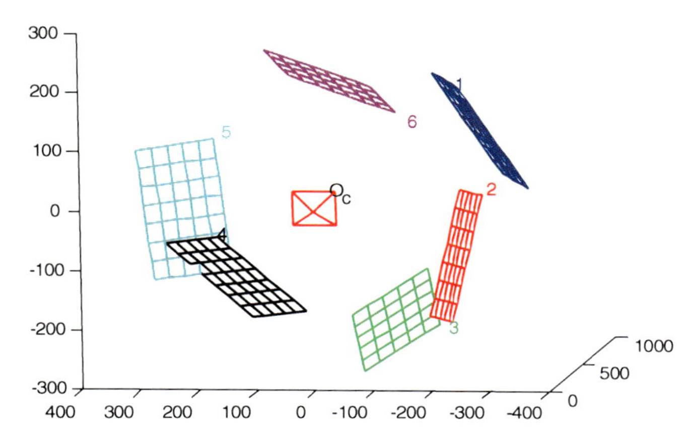

Fig. 6.18 The reconstructed positions of the pattern

Table 6.2 Back-projection errors in the image space (unit: pixel)

Items	U-component		V-component	
	Mean	Std Dev	Mean	Std Dev
1	0.5364	0.4113	0.8448	0.5701
2	0.5446	0.4038	1.0718	0.9099
3	0.5783	0.3651	0.6022	0.8175
4	0.6690	0.4956	0.7050	0.5234
5	0.7529	0.6596	0.9836	0.7769
6	1.1602	0.7797	0.5384	0.4908

In this case, we can formulate the calibration task into a polynomial eigenvalue problem.

6.3.5.1 Mirror-to-Mirror Homography

Given an arbitrary point $\boldsymbol{P}_o$ lies on a planar patch with the normal $\boldsymbol{n}$, we have $\boldsymbol{n}^{\mathrm{T}}\boldsymbol{P}_o = 1$. Combined with (6.13), we obtain

$$\boldsymbol{P}'_o = \boldsymbol{H}\boldsymbol{P}_o \tag{6.40}$$

where $\boldsymbol{H} = \boldsymbol{R} + \boldsymbol{t}\boldsymbol{n}^{\mathrm{T}}$ is a 3×3 matrix.

Considering (6.7), (6.9) and (6.40), we have

$$\boldsymbol{P}'_m = \sigma\boldsymbol{H}\boldsymbol{P}_m \tag{6.41}$$

where $\sigma = \lambda'/\lambda$ is a scale factor.

Equation (6.41) reveals the homogeneous transformation of the mirror surface before and after a rigid motion. Hence, the matrix $\boldsymbol{H}$ is named as mirror-to-mirror Homography.

6.3.5.2 Constraints and Solutions

Assuming $\boldsymbol{u}$ and $\boldsymbol{u}'$ are the image points of $\boldsymbol{P}_o$ before and after the rigid motion of $\boldsymbol{R}$ and $\boldsymbol{t}$. Let $\boldsymbol{u} = [u \quad v]^{\mathrm{T}}$ and $\boldsymbol{u}' = [u' \quad v']^{\mathrm{T}}$. According the back-projection in (6.10), we have

$$\begin{bmatrix} u' \\ v' \\ \frac{u'^2 + v'^2 - a^2}{2a} \end{bmatrix} = \sigma \boldsymbol{H} \begin{bmatrix} u \\ v \\ \frac{u^2 + v^2 - a^2}{2a} \end{bmatrix} \tag{6.42}$$

In this equation, the unknowns include the mirror parameter a, the Homography matrix $\boldsymbol{H}$ and the scale factor σ.

For simplicity, let $r = u^2 + v^2$ and $r' = u'^2 + v'^2$. Given a pair of image points, by eliminating σ, (6.42) can be rewritten as

$$(\boldsymbol{A}_0 + a\boldsymbol{A}_1 + a^2\boldsymbol{A}_2 + a^3\boldsymbol{A}_3 + a^4\boldsymbol{A}_4)\boldsymbol{h} = 0 \tag{6.43}$$

where

$$\boldsymbol{A}_0 = \begin{bmatrix} 0 & 0 & -rr' & 0 & 0 & 0 & 0 & 0 & 0 \\ 0 & 0 & 0 & 0 & 0 & -rr' & 0 & 0 & 0 \end{bmatrix},$$

$$\boldsymbol{A}_1 = \begin{bmatrix} -2r'u & -2r'v & 0 & 0 & 0 & 0 & 0 & 0 & 2u'r \\ 0 & 0 & 0 & -2r'u & -2r'v & 0 & 0 & 0 & 2v'r \end{bmatrix},$$

$$\boldsymbol{A}_2 = \begin{bmatrix} 0 & 0 & r'+r & 0 & 0 & 0 & 4u'u & 4u'v & 0 \\ 0 & 0 & 0 & 0 & 0 & r'+r & 4v'u & 4v'v & 0 \end{bmatrix},$$

$$\boldsymbol{A}_3 = \begin{bmatrix} 2u & 2v & 0 & 0 & 0 & 0 & 0 & 0 & -2u' \\ 0 & 0 & 0 & 2u & 2v & 0 & 0 & 0 & -2v' \end{bmatrix},$$

$$\boldsymbol{A}_4 = \begin{bmatrix} 0 & 0 & -1 & 0 & 0 & 0 & 0 & 0 & 0 \\ 0 & 0 & 0 & 0 & 0 & -1 & 0 & 0 & 0 \end{bmatrix} \text{ and}$$

$$\boldsymbol{h} = [h_1 \quad h_2 \quad h_3 \quad h_4 \quad h_5 \quad h_6 \quad h_7 \quad h_8 \quad h_9]^{\mathrm{T}}.$$

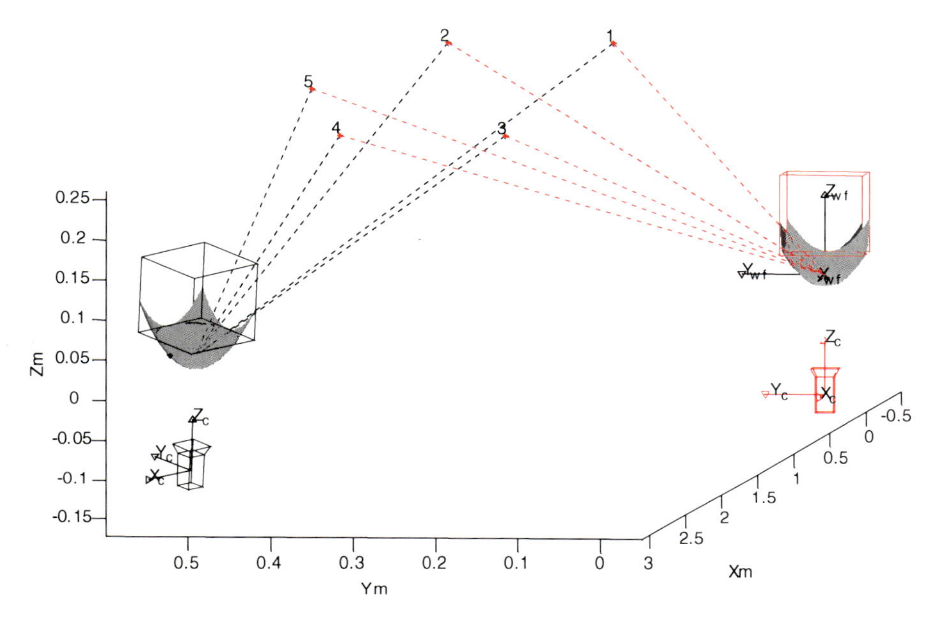

Fig. 6.19 Layout of the simulation system

From (6.43), we can see that each pair of image points contribute two rows into the coefficient matrices. Hence, five pairs are sufficient to solve that matrix equation. This is known as the fourth-order Polynomial Eigenvalue Problem. We can easily solve it by the function *polyeig* in MATLAB. The input is the designed matrices $\boldsymbol{A}_i$ which should be square matrix. The output consists of two elements. One is a 9×36 matrix whose columns are the eigenvectors, and the other is a vector of length 36, whose elements are the eigenvalues.

There are 36 solutions for a, many of them are zero, infinite or complex. In practice no more than 3 are considerable. The one which has the smallest sum of squared error for (6.42) is chosen. The solutions can be refined using levenberg-marquardt algorithm by minimizing

$$\min \left\|(\boldsymbol{A}_0 + a\boldsymbol{A}_1 + a^2\boldsymbol{A}_2 + a^3\boldsymbol{A}_3 + a^4\boldsymbol{A}_4)\boldsymbol{h}\right\| + \varepsilon(a^2 + \boldsymbol{h}^{\mathrm{T}}\boldsymbol{h})$$

where ε is a small penalty factor.

Once the matrix $\boldsymbol{H}$ is obtained, it can be decomposed into the motion parameters, i.e. the rotation matrix and translation vector. Finally, the calibration results can be optimized by bundle adjustment with all the available image points.

6.3.5.3 Test Example

We present a numerical test example here. Figure 6.19 illustrates a simulation system in the experiments which consists of a parabolic mirror and a conventional camera.

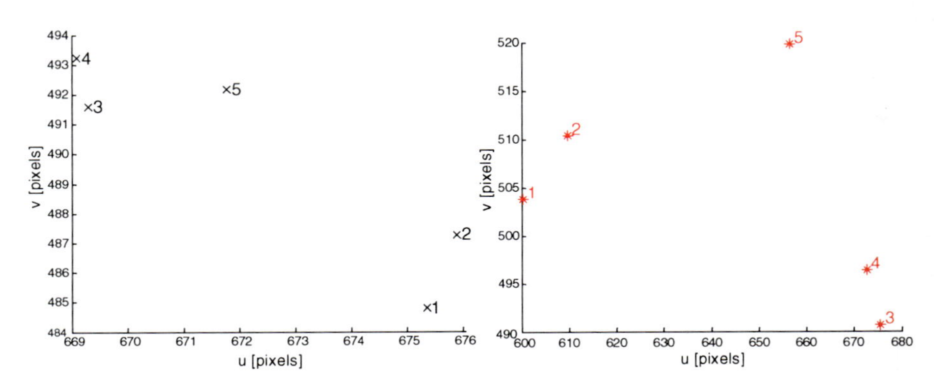

Fig. 6.20 Simulation images: the left and right figures respectively represent images before and after the motion, and each contains five corresponding pixels

The system undergoes an arbitrary rigid motion when working. The task is to estimate the motion parameters and the mirror parameter.

In this test example, we assumed that the system had the following parameters: the mirror $a = 0.03$, the rotation angles $(\pi/36, \pi/15, \pi/4)$ and translation vector $[3 \quad 0.5 \quad 0.05]$. The five randomly generated coplanar points were shown in Fig. 6.19 and their image points in Fig. 6.20. With the point pairs, we constructed the five matrices $\boldsymbol{A}_{0-4}$ in Eq. (6.39). Then we solved the polynomial eigenvalue problem to obtain an eigenvalue 0.03 as the mirror parameter and its corresponding eigenvector

$$\boldsymbol{h} = [0.0789, -0.0486, 0.9756, 0.0602, 0.0568, 0.1542, -0.0113, 0.0185, 0.0924]^{\mathrm{T}}$$

which represented the Homography.

With the known Homographic matrix $\boldsymbol{H}$, the rotation matrix and translation vector were obtained as what follows

$$\boldsymbol{R} = \begin{bmatrix} 0.6917 & -0.6709 & 0.2674 \\ 0.7071 & 0.7044 & -0.0616 \\ -0.1470 & 0.2317 & 0.9616 \end{bmatrix} \quad \text{and} \quad \boldsymbol{t} = [0.9863 \quad 0.1644 \quad 0.0164]^{\mathrm{T}}.$$

We can see that the estimated results are exactly equal to the assumed configuration.

6.4 Hyperbolic Camera System

6.4.1 System Structure

This section talks about the hyperbolic camera system which consists of a CCD camera and a hyperboloid mirror. Figure 6.21 shows a sketch of such system. The hyperboloid is a quadratic surface of revolution with the symmetry axis passing

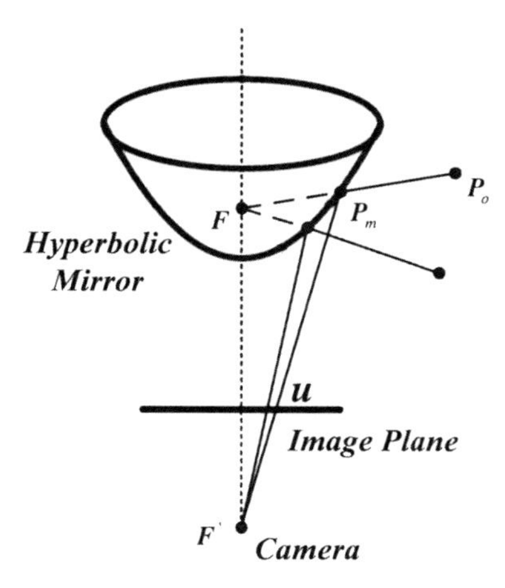

Fig. 6.21 A sketch of the hyperbolic camera system

through its two focal points, as denoted by $\boldsymbol{F}$ and $\boldsymbol{F}'$ respectively. According to Snell's law, all the 3D rays reflected by the mirror surface via one focal point should converge at the other. Hence, the system can be configured in a way that the optical center of the camera coincides with the second focal point while its optical axis aligns with the symmetric axis of the mirror. Such configuration ensures central projection process and considerably simplifies the imaging model.

6.4.2 Imaging Process and Back Projection

Let the origins of the camera and mirror coordinate frames locate at focal point $\boldsymbol{F}$, and the Z-axis coincides with the symmetry axis. Let e be the mirror eccentricity. Then the rotation matrix can be set $\boldsymbol{R}_c = \boldsymbol{I}$ and the translation vector $\boldsymbol{t}_c = [0\ \ 0\ \ 2e]^{\mathrm{T}}$.

Mathematically, the hyperbolic mirror is expressed as

$$\frac{(z+e)^2}{a^2} - \frac{x^2+y^2}{b^2} = 1 \tag{6.44}$$

where a and b are mirror parameters representing the semi-major and semi-minor axis respectively, and $e = \sqrt{a^2+b^2}$.

Similarly, the relationship between an arbitrary point $\boldsymbol{P}_o$ in the object and its corresponding point $\boldsymbol{P}_m$ on the mirror surface can be described by (6.7). Put it into (6.44) to determine the scale factor, we have two solutions and the needed one is $\lambda = \frac{b^2(-ze+a\sqrt{x^2+y^2+z^2})}{z^2b^2-(x^2+y^2)a^2}$. And so is the point $\boldsymbol{P}_m$.

Then the perspective projection from $\boldsymbol{P}_m$ to the image point $\boldsymbol{u}$ can be written as

$$\boldsymbol{u} = \begin{bmatrix} \frac{p_1}{(p_3+2e)} \\ \frac{p_2}{(p_3+2e)} \\ 1 \end{bmatrix} \tag{6.45}$$

where $\boldsymbol{P}_i$ represents i-th element in $\boldsymbol{P}_m$.

When the concerned object performs a rigid motion, the point $\boldsymbol{P}_o$ is moved to the position of $\boldsymbol{p}_o'$ and we can have another image point, say $\boldsymbol{u}'$.

Now supposing we have an image point $\boldsymbol{u}$, we consider the back-projection process. From Fig. 6.21, it is obvious that the three points, i.e. $\boldsymbol{F}'$, $\boldsymbol{u}$ and $\boldsymbol{P}_m$, are collinear, which means

$$\boldsymbol{p}_m = \boldsymbol{t}_c + \lambda \boldsymbol{u} \tag{6.46}$$

Substituting (6.46) into (6.44), the scale factor λ can be solved. Although there are two solutions for λ, the correct one can be easily found considering the position relationship of the three points. So are the point $\boldsymbol{P}_m$ and the 3D ray connecting the two points $\boldsymbol{P}_m$ and $\boldsymbol{P}_o$. Consequently, 3D reconstruction can be implemented if two or more such rays are obtained.

6.4.3 Polynomial Eigenvalue Problem

Considering the point $\boldsymbol{P}_o$ on a planar patch, its corresponding point on the mirror surface before and after rigid motion still verifies Eq. (6.41). The variable parameters include the motion parameters and the two mirror parameters, i.e. a and b. We will show how to formulate them into polynomial eigenvalue problem.

The mirror point $\boldsymbol{P}_m$ in Eq. (6.46) can be considered as a function of the mirror parameters, which means

$$\boldsymbol{P}_m = f(a, b) \tag{6.47}$$

Applying the first order Taylor series expansion to (6.47) at $a = a_0$ and $b = b_0$, we have

$$\boldsymbol{P}_m = f(a_0, b_0) + (a - a_0) f_a(a_0, b_0) + (b - b_0) f_b(a_0, b_0) \tag{6.48}$$

where $f_a(a_0, b_0)$ and $f_b(a_0, b_0)$ denote partial derivatives of $f(a, b)$ with respect to a and b. In MATLAB, their expressions can be derived by the commands $jacobian(f(a,\ b),\ a)$ and $jacobian(f(a,\ b),\ b)$, respectively. Due to limited space, we omit their expressions.

Rearranging the above equation, we obtain

$$\boldsymbol{P}_m = \boldsymbol{k} + a\boldsymbol{a} + b\boldsymbol{b} \tag{6.49}$$

where $\boldsymbol{k} = f(a_0, b_0) - a_0 f_a(a_0, b_0) - b_0 f_b(a_0, b_0)$, $\boldsymbol{a} = f_a(a_0, b_0)$ and $\boldsymbol{b} = f_b(a_0, b_0)$.

Similarly, the corresponding point on the mirror surface after motion has the following formula

$$\boldsymbol{p}' = \boldsymbol{k}' + a\boldsymbol{a}' + b\boldsymbol{b}' \tag{6.50}$$

Combining (6.41), (6.49) and (6.50), we have

$$\boldsymbol{k}' + a\boldsymbol{a}' + b\boldsymbol{b}' = \sigma \boldsymbol{H}(\boldsymbol{k} + a\boldsymbol{a} + b\boldsymbol{b}) \tag{6.51}$$

In (6.51), the unknowns include the mirror parameter a and b, the Homography matrix $\boldsymbol{H}$ and the scale factor $\boldsymbol{\sigma}$. The values of $\boldsymbol{k}$, $\boldsymbol{a}$, $\boldsymbol{b}$ and $\boldsymbol{k}'$, $\boldsymbol{a}'$, $\boldsymbol{b}'$ can be computed from the image point correspondences. By eliminating $\boldsymbol{\sigma}$, we have the following constraint

$$(\boldsymbol{A}_0 + a\boldsymbol{A}_1 + a^2\boldsymbol{A}_2)\boldsymbol{h} = 0 \tag{6.52}$$

where $\boldsymbol{h} = [h_1 \quad h_2 \quad \cdots \quad h_9 \quad bh_1 \quad bh_2 \quad \cdots \quad bh_9 \quad b^2h_1 \quad b^2h_2 \quad \cdots \quad b^2h_9]^{\mathrm{T}}$, $\boldsymbol{A}_0$, $\boldsymbol{A}_1$ and $\boldsymbol{A}_2$ are square matrices and can be constructed from the image point correspondences as what we do in the previous section. We can see that (6.52) is a polynomial eigenvalue problem with degrees 2.

Similarly, each correspondence contributes two rows for the three coefficient matrices. Hence, fourteen correspondences are sufficient to solve it. The parameter a is directly obtained from the eigenvalue while parameter b and the motion parameters can be extracted from the corresponding vector $\boldsymbol{h}$.

6.5 Summary

This chapter talks about recent progress in the catadioptric vision system, including its background, its configuration and calibration problems. Due to the specific optics structures, some distinct geometric properties have been found within such kind of system.

Three different systems are mainly discussed, i.e. panoramic stereoscopic system, parabolic camera system and hyperbolic camera system. The properties of cataoptric via mirror surface and dioptric via lens are preserved in those systems. They all feature a large field of view and central projection with a single effective viewpoint, which are important in vision tasks. The first system consists of a hyperbolic mirror and a CCD camera with fisheye lens. In this system, the overlapping field of view of the fisheye image and catadioptric image provide stereo information for 3D reconstruction over a 360 horizontal field of view. A generic radial distortion model is used to model the imaging process, which can easily accommodate the non-single viewpoint issue in the catadioptric image. The second system combines a parabolic mirror and a camera with orthographic projection. When calibration its parameters, both lifted fundamental matrix and homographic matrix are employed. Here, both general motion case and specific motion case are discussed. The third system is composed by a hyperbolic mirror and a conventional perspective camera. Compared with the first one, the imaging geometry in the second and third systems is simpler and can be modeled in an analytic form. Finally, some results of numerical simulations and real data experiments are presented to show the validity of those algorithms for the vision systems.

The wide field of view enables the catadioptric vision system highly applicable for many applications since more information can be inferred from fewer images in such system. In practice, it is not often required to capture the entire spherical field of view. In most cases, it is sufficient to capture 360° panoramas with a reasonable vertical field of view that has more or less equal coverage above and beneath the horizon. For instance, in the case of video conferencing, we only need to capture the sitting participants and the tables. The same is true for navigation, where the vehicle only need to view the roads and the obstacles around it. Hence, some compromises among the computational complexity, robustness and viewing angle should be carefully made when bringing the catadioptric vision system from laboratory to real world use.

Chapter 7
Conclusions and Future Expectation

This chapter provides a brief summary of the book and outlines the possible problems for future investigation.

7.1 Conclusions

In this book, the relevant issues of static calibration, dynamic calibration and 3D reconstruction in the context of various active vision systems are investigated. The main work can be summarized as what follows.

The task of static calibration is to estimate the fixed intrinsic parameters, such as the focal lengths, the aspect ratio and principal point, etc. The theories of camera calibration are discussed and necessary constraints are established via the image of absolute conic. Three different kinds of planar patterns are designed for implementing the calibration task. The first pattern contains an equilateral polygon with $2n$ sides ($n > 2$), in which the projections of vertexes are used to approximate the ellipse and those edges for computing the vanishing line. The second pattern consists of two intersecting circles, from which the projections of circular points can be directly obtained via the solutions of quartic equations. The third pattern composes two concentric circles. The approach for determining the projection of the circle center is investigated with the pole-polar relationship. Then the solution is formulated into a first-order polynomial eigenvalue problem. It is obvious that all of the calibration objects in these patterns have simple structures, and there are no limitations on their positions and metric sizes. Hence, it is easy to prepare such patterns. Another characteristic is that they not only provide sufficient geometric constraints but also are effortlessly detectable and identifiable, requiring no point correspondences. Therefore, it is convenient to implement those algorithms and the computational complexity is efficient.

The dynamic calibration is to determine the possible variable parameters in the vision system, including intrinsic and extrinsic parameters. The plane-based Homography and fundamental matrix as well as the epipolar geometry are investigated

B. Zhang, Y. F. Li, *Automatic Calibration and Reconstruction for Active Vision Systems*, Intelligent Systems, Control and Automation: Science and Engineering,
DOI 10.1007/978-94-007-2654-3_7, © Springer Science+Business Media B.V. 2012

to accomplish this task. The eigenvectors and eigenvalues of the homogrpahy matrix are extensively analyzed to provide constraints for the motion parameters in the structured light system and the catadioptric system. We also show that the calibration problem can be converted into a polynomial eigenvalue problem through the homography matrix. On the other hand, auto-epipolar relationship can be established in a panoramic stereoscopic system, which greatly simplifies its geometric model and improves the robustness of computation. With the lifted coordinates in a parabolic camera system, further use of fundamental matrix is discussed, in which a 4×4 matrix with rank 2 is defined. Linear algorithms with closed-form solutions are explored to decrease the computational load and increase the calibration accuracy. In the presence of noise, redundancy in the calibration data can be easily incorporated to enhance the robustness. Hence, those techniques are highly reliable and pretty applicable for dynamic environment with moving or deformable objects.

The image-to-world transformation is investigated in obtaining 3D Euclidean structure of the scene in the structured light system. Two different cases are considered, i.e. two-known-plane case and one-known-plane case. Here, the word 'known' means the equation of a plane is known, rather than knowing any structure in the plane. In the first case, we show that dynamic determination of the transformation is equivalent to the computation of online image-to-light-plane Homography. In the second case, the varying focal lengths and extrinsic parameters of the camera are firstly calibrated using the scene-plane-to-image Homography. Then the image-to-light-plane Homography can be easily determined, and so is the image-to-world transformation. In addition, the reconstruction algorithm is proved in both theory and simulation experiments to be computationally cheaper than the traditional triangulation approaches.

Different imaging models of the vision system are established in the context of different environments. For example, the projector in the structured light system can be considered as a collection of light planes and the system is modeled as those planes with the camera. Then in Chap. 4, the projector is deemed as a virtual camera and the system as an active stereoscopic vision. In the catadioptric system, three possible combinations of lenses and mirrors are tested. Accordingly, panoramic stereoscopic system, parabolic camera system and hyperbolic camera system are respectively constructed. They all have a single effective viewpoint with a large field of view.

Compared with other 3D techniques, e.g. stereo vision or laser range finder, the structured light system features good accuracy, high reliability, and fast data acquisition speed. However, there exist several limitations, as have been discussed in Sect. 5.5. Here, we should also be aware that the structured light system has the limitation of field of view and depth of field. The interested features of the object should fall into the field of views of both the camera and the projector. For a typical camera with a 2/3″ sensor and a 25 mm lens, the field of view only has an angle just over 10°. Thus one action of 3D acquisition can only obtain a small amount of surface area. The depth of field of a camera is also very limited. For such a typical camera, the object is best placed in a depth range of about 120 mm. Out of that range will cause significant decrease of the accuracy because of image blurring. The catadioptric system overcomes those limitations by using a specially-shaped mirror, whose horizontal angle of view can reach as large as 360°. However, severe distortion exists in its image which brings a lot of troubles in the image processing.

7.2 Future Expectations

Following the existing work discussed in this book, some possible expectations are suggested:

Without Knowledge from the Scene In Chap. 5, the structured light system is considered as a collection of light planes and a camera. When some parameters are adjusted, the image-to-light-plane Homography $\boldsymbol{H}_k$ is changed. In this case, how to compute $\boldsymbol{H}_k$ dynamically is a key problem for 3D reconstruction by the image-to-world transformation. Two different cases have been discussed, i.e. two-known-plane case and one-known-plane case. The readers may also be interested in the case of a completely arbitrary scene. Such problem can be tackled as what follows.

For an arbitrary point $\tilde{\boldsymbol{M}}_{ki}$ on the k-th light plane, we have $\boldsymbol{n}_k^{\mathrm{T}}\tilde{\boldsymbol{M}}_{ki} = 0$.

Let $\tilde{\boldsymbol{m}}_i$ be the image point of $\tilde{\boldsymbol{M}}_{ki}$. From (5–10), we obtain $\boldsymbol{n}_k^{\mathrm{T}}\boldsymbol{T}_k^{-1}\boldsymbol{S}^{\mathrm{T}}\boldsymbol{H}_k\tilde{\boldsymbol{m}}_i = 0$.

This provides one constraint on the Homographic matrix $\boldsymbol{H}_k$. However, the experiments show that $\boldsymbol{H}_k$ cannot be fully determined no matter how many points are used. If another arbitrary camera is employed simultaneously, three Homographic matrices arise for each light plane and obviously they are not independent with each other. To investigate their relationships so as to enable the determination of $\boldsymbol{H}_k$ may be an interesting work.

Integration of Methodologies Integration of the structured light techniques and shape from shading techniques may be a useful try. The idea is from the fact that the structured light projection provides intentional light source while the assumptions of knowing light source direction or a single point light source are usually used in the shape from shading techniques. The intensity of a point in the image constrains the image irradiance with the diameter of camera's aperture, the focal lengths, the distance between the projector and the object surface, the angle between the surface normal and incident ray, the angle between the specular reflection direction and observing direction. By analysis of the intensity distribution on an image, we may find the cues among the object depth, camera pose, and light source pose. Such cues then can be used to determine the intrinsic or extrinsic parameters of the system. Derivation of such formulations can be considered as a part of the future work.

Self-adaptive Vision System The behavior of the color-encoded light pattern should be studied when it is projected onto a scene with color surface. Relevant information may be utilized to refine the vision system. Obviously, a frequency shifting happens to the color captured by the camera with respect to the color projected by the light pattern due to the intrinsic color of the surface. If the colors in the pattern are well chosen, this discrepancy may allow us to obtain the intrinsic color of the surface. Then the system could be used to obtain 3D structure information and color information of any scene by a single pattern shot projection. Furthermore, the obtained information in the current frame provides a strong constraint on the next snapshot, which in turn may be used to construct a self-adaptive vision system.

Catadioptric Projector System In general, severe distortions such as stretching and shearing distortion, etc, are observed in this kind of system. Hence, how to significantly and efficiently reduce both the geometric and photometric distortions is a preliminary problem. Anyway, integration the merits from both catadioptric vision system and structured light system is another interesting expectation. Recently, some researchers have made attempts following this direction. For example, in [214], the authors described a catadioptric omnidirectional video-projection system which can adjust its projection to the geometry of any scene by means of a rotating camera. The configuration of this system consists of a projector, a camera and a hemispherical mirror, in which the camera is used to detect projected point features. The calibration is performed in three successive steps: precalibration of camera assuming pure rotation, precalibration of catadioptric projector under central approximation and calibration of the global system, by minimizing the squared distance between the reflected and perceived rays. In [215], the authors discussed efficient algorithms to reduce projection artifacts such as distortions, scattering, and defocusing in the catadioptric projector system. The light transport matrix is employed to model the light transport between the projector and the viewpoint. Although it is very difficult to determine the light transport matrix in practice, those results provide a distinct conceptual inspiration for designing new generations of vision systems. Hence, the future work may include how to efficiently estimate the light transport matrix and develop real-time distortion and defocusing compensation methods.

References

1. Roberts LG (1965) Machine perception of three-dimensional solids, optical and electro-optical. MIT Press, Cambridge, pp 159–197
2. Marr D (1982) Vision: a computational investigation into the human representation and processing of visual information. W.H. Freeman, New York
3. Faugeras O (1992) What can be seen in three dimensions with an uncalibrated stereo rig? Proceedings of the European conference on computer vision (ECCV'92), LNCS-Series, vol 588. Springer, New York, pp 563–578, 1992
4. Heyden A (1995) Reconstruction from image sequences by means of relative depths. Proceedings of the IEEE international conference on computer vision, pp 1058–1063, 1995
5. Beardsley P, Zisserman A, Murray D (1997) Sequential update of projective and affine structure from motion. Int J Comput Vis 23(3):235–259
6. Avidan S, Shashua A (1998) Threading fundamental matrices. Proceedings of the European conference on computer vision, University of Freiburg, Germany, pp 124–140
7. Shashua A (1995) Multiple-view geometry and photometry. Asian conference on computer vision (ACCV95), Singapore, pp 395–404, Dec 1995
8. Shashua A (1997) On photometric issues in 3D visual recognition from a single 2D image. Int J Comput Vis 21(1):99–122
9. Hartley R (1997) Lines and points in three views and the trifocal tensor. Int J Comput Vis 2(2):125–140
10. Tomasi C, Kanade T (1992) Shape and motion form image streams under orthography: a factorization approach. Int J Comput Vis 9(2):137–154 (Nov 1992)
11. Reid I, Murray D (1996) Active tracking of foveated feature clusters using affine structure. Int J Comput Vis 18(1):41–60 (Apr 1996)
12. Rother C, Carlsson S (2001) Linear multi-view reconstruction and camera recovery. IEEE international conference on computer vision, Vancouver, Canada, pp 42–51
13. Rother C (2003) Multi-view reconstruction and camera recovery using a real or virtual reference plane, PhD thesis
14. Sturm P, Triggs B (1996) A factorization based algorithm for multi-image projective structure and motion. Proceedings of the European conference on computer vision, Cambridge, England, pp 709–720, Apr 1996
15. Han M, Kanade T (2001) Multiple motion scene reconstruction from uncalibrated views, IEEE international conference on computer vision, vol 1, pp 163–170
16. Hung Y, Tang W (2006) Projective reconstruction from multiple views with minimization of 2D reprojection error. Int J Comput Vis 66(3):305–317
17. Anas H, Fisker R et al (2002) Robust factorization. IEEE Trans Pattern Anal Mach Intell 24(9):1215–1225
18. Huynh D, Hartley R, Heyden A (2003) Outlier correction in image sequences for the affine camera. IEEE international conference on computer vision, vol 1, Nice, France, pp 585–590, 11–17 Oct 2003

B. Zhang, Y. F. Li, *Automatic Calibration and Reconstruction for Active Vision Systems*, Intelligent Systems, Control and Automation: Science and Engineering, DOI 10.1007/978-94-007-2654-3,

19. Brandt S (2005) Conditional solutions for the affine reconstruction of N-views. Image Vis Comput 23:619–630
20. Hartley R (1993) Euclidean reconstruction from uncalibrated views. Lecture notes in computer science, Springer, Berlin, pp 237–256
21. Heyden A, Astrom K (1996) Euclidean reconstruction from constant intrinsic parameters. Proceedings of the international conference on pattern recognition, vol 1, Vienna, pp 339–343, 25–29 Aug 1996
22. Pollefeys M, Van Gool L, Osterlinck A (1996) The modulus constraint: a new constraint for self-calibration. Proceedings of the international conference on pattern recognition, vol 1, Vienna, pp 349–353, 25–29 Aug 1996
23. Faugeras O (1995) Stratification of three-dimensional projective, affine and metric representations. J Opt Soc Am 12(3):465–484
24. Huynh D, Heyden A (2005) Scene point constraints in camera auto-calibration: an implementational perspective. Image Vis Comput 23(8):747–760
25. Gong R, Xu G (2004) 3D structure from a single calibrated view using distance constraints. Proceedings Asian conference on computer vision, vol 1, pp 372–377
26. Habed A, Boufama B (2006) Camera self-calibration form bivariate polynomial equations and the coplanarity constraint. Image Vis Comput 24(5):498–514
27. Liebowitz D, Zisserman A (1999) Combining scene and auto-calibration constraints. IEEE international conference on computer vision, vol 1, pp 293–300, Sept 1999
28. Heyden A, Huynh D (2002) Auto-calibration via the absolute quadric and scene constraints. International conference on pattern recognition, vol 2. Quebec, Canada, pp 631–634, Aug 2002
29. Hartley R, Hayman E et al (1999) Camera calibration and the search for infinity. IEEE international conference on computer vision, vol 1, pp 510–517, Sept 1999
30. Tsai R, Versatile A (1987) Camera calibration technique for high-accuracy 3D machine vision metrology using off-the-shelf TV cameras and lenses. IEEE J Robot Autom 3(4):323–344 (Aug 1987)
31. Hu R, Ji Q (2001) Camera self-calibration from ellipse correspondences. Proceedings IEEE international conference on robotics and automation, Seoul, Korea, pp 2191–2196, 21–26 May 2001
32. Kaminiski Y, Shashua A (2004) On calibration and reconstruction from planar curves. Int J Comput Vis 56(3):195–219 (Feb–March 2004)
33. Wilczkowiak M, Sturm P, Boyer E (2005) Using geometric constraints through parallelepipeds for calibration and 3D modeling. IEEE Trans Pattern Anal Mach Intell 27(2):194–207
34. Zhang Z (2000) A flexible new technique for camera calibration. IEEE Trans Pattern Anal Mach Intell 22(11):1330–1334
35. Sturm P, Maybank S (1999) On plane-based camera calibration: a general algorithm, singularities, applications. Proceedings of the IEEE conference on computer vision and pattern recognition, Fort Collins, USA, pp 432–437
36. Ueshiba T, Tomita F (2003) Plane-based calibration algorithm for multi-camera systems via factorization of homography matrices. IEEE international conference on computer vision, pp 966–973
37. Shashua A, Avidan S (1996) The rank 4 constraint in multiple (> = 3) view Geometry. European conference on computer vision, pp 196–206
38. Zelnik-Manor L, Irani M (2005) Multiview constraints on homographies. IEEE Trans Pattern Anal Mach Intell 24(2):214–223
39. Malis E, Cipolla R (2002) Camera self-calibration from unknown planar structures enforcing the multiview constraints between collineations. IEEE Trans Pattern Anal Mach Intell 24(9):1268–1272
40. Maybank S, Faugeras O (1992) A theory of self-calibration of a moving camera. Int J Comput Vis 8(2):123–151
41. Pollefeys M, Kock R, Van Gool L (1998) Self-calibration and metric reconstruction in spite of varying and unknown internal camera parameters, IEEE international conference on computer vision, pp 90–95

42. Triggs B (1997) Autocalibration and the absolute quadric. Proceedings of the IEEE conference on computer vision and pattern recognition, pp 607–614
43. Hassanpour R, Atalay V (2004) Camera auto-calibration using a sequence of 2D images with small rotations. Pattern Recognit Lett 25:989–997
44. Huang C, Chen C, Chung P (2004) An improved algorithm for two-image camera self-calibration and Euclidean structure recovery using absolute quadric. Pattern Recognit 37(8):1713–1722
45. Ponce J, McHenry K et al (2005) On the absolute quadratic complex and its application to autocalibration. Proceedings of the IEEE conference on computer vision and pattern recognition, pp 780–787
46. Hartley R (1997) Kruppa's equations derived from the fundamental matrix. IEEE Trans Pattern Anal Mach Intell 19(2):133–135
47. Manolis I, Deriche R (2000) Camera self-calibration using the Kruppa equations and the SVD of the fundamental matrix: the case of varying intrinsic parameters, INRIA, No. 3911, Mars
48. Faugeras O, Luong Q, Maybank S (1992) Camera self-calibration: theory and experiments. Proceedings of ECCV. Springer, Heidelberg, pp 321–334
49. Sturm P (2000) A case against Kruppa's equations for camera self-calibration. IEEE Trans Pattern Anal Mach Intell 22(10):1199–1204
50. Zhang Z, Hanson A (1996) 3D reconstruction based on homography mapping. ARPA image understanding workshop, Palm Springs, CA
51. Liebowitz D, Zisserman A (1998) Metric rectification for perspective images of planes. Proceedings of IEEE conference on computer vision and pattern recognition, Santa Barbara, California, pp 482–488, June 1998
52. Knight J, Zisserman A, Reid I (2003) Linear auto-calibration for ground plane motion. Proceedings of IEEE conference on computer vision and pattern recognition, pp 503–510
53. Quan L, Wei Y et al (2004) Constrained planar motion analysis by decomposition. Image Vis Comput 22:379–389
54. Gurdjos P, Crouzil A, Payrissat R (2002) Another way of looking at plane-based calibration: the centre circle constraint. European conference on computer vision, pp 252–266
55. Bocquillon B, Gurdjos P, Crouzil A (2006) Towards a guaranteed solution to plane-based self-calibration. Proceedings of the 7th Asian conference on computer vision, vol 1, pp 11–20
56. Gurdjos P, Sturm P (2003) Methods and geometry for plane-based self-calibration. Proceedings of the international conference on computer vision and pattern recognition, pp 491–496
57. Kim J, Gurdjos P, Kweon I (2005) Geometric and algebraic constraints of projected concentric circles and their applications to camera calibration. IEEE Trans Pattern Anal Mach Intell 27(4):1–6
58. Fusiello A (2000) Uncalibrated Euclidean reconstruction: a review. Image Vis Comput 18:555–563
59. Lu Y, Jason Z, Zhang et al (2004) A survey of motion-parallax-based 3-D reconstruction algorithms. IEEE Trans Syst Man Cybern Part C: Appl Rev 34(4):1–17 (Nov 2004)
60. Bowyer W, Chang K, Flynn P (2006) A survey of approaches and challenges in 3D and multi-modal 3D + 2D face recognition. Comput Vis Image Underst 101(1):1–15
61. Zhong H, Hung YS (2006) Self-calibration from one circular motion sequence and two images. Pattern Recognit 39(9):1672–1678
62. Scharstein D, Szeliski R (2003) High-accuracy stereo depth maps using structured light. IEEE computer society conference on computer vision and pattern recognition (CVPR 2003), Madison, pp 195–202, June 2003
63. Zhang Li, Curless B, Seitz M (2002) Rapid shape acquisition using color structured light and multi-pass dynamic programming. Proceedings of the 1st international symposium on 3D data processing, visualization, and transmission (3DPVT), Padova, Italy, pp 24–36, 19–21 June 2002
64. Zhang Li, Curless B, Seitz M (2003) Spacetime stereo: shape recovery for dynamic scenes. Proceedings of IEEE computer society conference on computer vision and pattern recognition (CVPR), Madison, pp 367–374, June 2003

65. Davis J, Ramamoorthi R, Rusinkiewicz S (2005) Spacetime stereo: a unifying framework for depth from triangulation. IEEE Trans Pattern Anal Mach Intell 27(2):296–302
66. Chu CW, Hwang S, Jung SK (2001) Calibration-free approach to 3D reconstruction using light stripe projections on a cube frame. IEEE 3rd international conference on 3D digital imaging and modeling, Quebec, Canada, pp 13–19, 28 May–01 June 2001
67. Chen CH, Kak AC (1987) Modeling and calibration of a structured light scanner for 3-D robot vision. Proceedings IEEE conference on robotics and automation, pp 807–815
68. Reid ID (1996) Projective calibration of a laser-stripe range finder. Image Vis Comput 14:659–666
69. Huynh D, Owens R, Hartmann P (1999) Calibrating a structured light stripe system: a novel approach. Int J Comput Vis 33:73–86
70. DePiero FW, Trivedi MM (1996) 3D Computer Vision Using Structured Light: design, calibration and implementation issues. Adv Comput 43:243–278
71. Sansoni G, Carocci M, Rodella R (2000) Calibration and performance evaluation of a 3D imaging sensor based on the projection of structured light. IEEE Trans Instrum Meas 49(3):628–636 (June 2000)
72. McIvor AM (2002) Nonlinear calibration of a laser stripe profiler. Opt Eng 41(1):205–212
73. Bouguet JY, Perona P (1998) 3D photography on your desk. IEEE international conference on computer vision, Bombay, India, pp 43–50
74. Fisher RB, Ashbrook AP, Robertson C, Werghi N (1999) A low-cost range finder using a visually located, Structured light source, 2nd international conference on 3D digital imaging and modeling, Ottawa, Canada, pp 24–33, 04–08 Oct 1999
75. Marzani S, Voisin Y et al (2002) Calibration of a three-dimensional reconstruction system using a structured light source. Opt Eng 41(2):484–492
76. Dipanda A, Woo S (2005) Towards a real-time 3D shape reconstruction using a structured light system. Pattern Recognit 38(10):1632–1650
77. Chen F, Brown GM, Song M (2000) Overview of three-dimensional shape measurement using optical methods. Opt Eng 39(1):10–22
78. Langer D, Mettenleiter M et al (2000) Imaging laser scanners for 3D modeling and surveying applications. Proceedings of the 2000 IEEE international conference on robotics and automation, San Francisco, pp 116–121, April 2000
79. Hebert P (2001) A self-referenced hand-held range sensor. Proceedings IEEE 3rd international conference on 3D digital imaging and modeling, Quebec, Canada, pp 5–12, May 2001
80. Fofi D, Salvi J, Mouaddib E (2001) Uncalibrated vision based on structured light. IEEE international conference on robotics and automation, Seoul, Korea, vol 4, pp 3548–3553, May 2001
81. Jokinen O (1999) Self-calibration of a light striping system by matching multiple 3D profile maps. Proceedings IEEE 2nd international conference on 3D digital imaging and modeling, Ottawa, Canada, pp 180–190
82. Wolfe W, Mathis D et al (1991)The perspective view of three points. IEEE Trans Pattern Anal Mach Intell 13(1):66–73
83. Haralick R, Lee C et al (1991) Analysis and solutions of the three point perspective pose estimation problem. Proceedings IEEE conference on computer vision and pattern recognition, Maui, Hawaii, pp 592–598
84. Horaud R, Conio B, Leboulleux O (1989) An analytic solution for the perspective 4-point problem. Comput Vis Graph Image Process 47(33–34):33–43
85. Quan L, Lan Z (1999) Linear N-point camera pose determination. IEEE Trans Pattern Anal Mach Intell 21(8):774–780
86. Triggs B (1999) Camera pose and calibration from 4 or 5 known 3D points. Proceedings of the 7th international conference on computer vision, Corfu, Greece, pp 278–284, Sept 1999
87. Abdel-Aziz Y, Karara H (1971) Direct linear transformation into object space coordinates in close-range photogrammetry. Proceedings symposium close-range photogrammetry, pp 1–18

88. Ganapathy S (1984) Decomposition of transformation matrices for robot vision. Proceedings IEEE conference robotics and automation, pp 130–139
89. Wang Z, Jepson A (1994) A new closed-form solution for absolute orientation. IEEE conference computer vision and pattern recognition (CVPR), pp 129–134, 21–23 June 1994
90. Fiore D (2001) Efficient linear solution of exterior orientation. IEEE Trans Pattern Anal Mach Intell 23(2):140–148
91. Wang F (2004) A simple and analytical procedure for calibrating extrinsic camera parameters. IEEE Trans Rob Autom 20(1):121–124
92. Wang F (2005) An efficient coordinate frame calibration method for 3D measurement by multiple camera systems. IEEE Trans Syst Man Cybern Part C 35(4):453–464, (Nov 2005)
93. Ansar A, Daniilidis K (2003) Linear pose estimation from points or lines. IEEE Trans Pattern Anal Mach Intell 25(5):578–588
94. Liu Y, Huang S, Faugeras O (1990) Determination of camera location from 2D to 3D line and point correspondences. IEEE Trans Pattern Anal Mach Intell 12(1):28–37
95. Liu Y, Zhang X, Huang S (2003) Estimation of 3D structure and motion from image corners. Pattern Recognit 36:1269–1277
96. Triggs B, McLauchlan P, Hartley R, Fitzgibbon A (2000) Bundle Adjustment—a modern synthesis. Springer lecture notes on computer science, vol 1983. Springer, Heidelberg, pp 298–375
97. Taylor C, Kriegman D (1995) Structure and motion from line segments in multiple images. IEEE Trans Pattern Anal Mach Intell 17(11):1021–1032
98. Oberkampf D, DeMenthon D, Davis S (1996) Iterative pose estimation using coplanar feature points. Comput Vis Image Underst 63(3):495–511 (May 1996)
99. Christy S, Horaud R (1999) Iterative pose computation from line correspondences. Comput Vis Image Underst 73(1):137–144
100. Lu C, Hager D, Mjolsness E (2000) Fast and globally convergent pose estimation from video images. IEEE Trans Pattern Anal Mach Intell 22(6):610–622
101. Koninckx T, Peers P, Dutre P, Van Gool L (2005) Scene-adapted structured light. IEEE computer society international conference on computer vision and pattern recognition—CVPR05, San Diego, USA, pp 611–619, 20–25 June 2005
102. Koninckx T, Van Gool L (2006) Real-time range acquisition by adaptive structured light. IEEE Trans Pattern Anal Mach Intell 28(3):432–445
103. Longuet-Higgins HC (1981) A computer algorithm for reconstructing a scene from two projections. Nature 293(9):133–135
104. Gruen A, Huang T (2001) Calibration and orientation of cameras in computer vision. Springer, New York
105. Faugeras O, Maybank S (1990) Motion from point matches: multiplicity of solutions. Int J Comput Vis 4:225–246
106. Hartley R (1992) Estimation of relative camera positions for uncalibrated cameras, lecture notes in computer science, ECCV'92, vol 588. Springer, Berlin, pp 579–587
107. Kanatani K, Matsunaga C (2000) Closed-form expression for focal lengths from the fundamental matrix. Proceedings 4th Asian conference on computer vision, pp 128–133, Jan 2000
108. Ueshiba T, Tomita F (2003) Self-calibration from two perspective views under various conditions: Closed-form solutions and degenerate configurations. Australia-Japan advanced workshop on computer vision, Sept 2003
109. Triggs B (2000) Routines for relative pose of two calibrated cameras from 5 points. Technical report, INRIA, France. http://www.inrialpes.fr/movi/people/Triggs
110. Nister D (2004) An efficient solution to the five-point relative pose problem. IEEE Trans Pattern Anal Mach Intell 26(6):756–770
111. Stewenius H, Engels C, Nister D (2006) Recent developments on direct relative orientation. J Photogramm Remote Sens 60(4):284–294 (June 2006)
112. Philip J (1996) A non-iterative algorithm for determining all essential matrices corresponding to five point pairs. Photogramm Rec 15(88):589–599 (Oct 1988)

113. Philip J (1998) Critical point configurations of the 5-, 6-, 7-, and 8-point algorithms for relative orientation. TRITA-MAT-1998-MA-13, Feb 1998
114. Hartley R (1997) In defense of the eight-point algorithm. IEEE Trans Pattern Anal Mach Intell 19(6):580–592
115. Wu FC, Hu ZY, Duan FQ (2005) 8-point algorithm revisited: factorized 8-point algorithm, ICCV05, Beijing, China, pp 488–494
116. Zhong HX, Pang YJ, Feng YP (2006) A new approach to estimating fundamental matrix. Image Vis Comput 24(1):56–60
117. Zhang Z (1998) Determining the epipolar geometry and its uncertainty: a review. Int J Comput Vis 27(2):161–198
118. Armangue X, Salvi J (2003) Overall view regarding fundamental matrix estimation. Image Vis Comput 21:205–220
119. Hay JC (1966) Optical motion and space perception: an extension of Gibson's analysis. Psychol Review 73(6):550–565
120. Tsai R, Huang T (1981) Estimating three dimensional motion parameters of a rigid planar patch. IEEE Trans Acoust Speech Signal Process ASSP-29:525–534
121. Tsai R, Huang T, Zhu W (1982) Estimating three dimensional motion parameters of a rigid planar patch, II: singular value decomposition. IEEE Trans Acoust Speech Signal Process ASSP-30:525–534
122. Longuet-Higgins HC (1984) The visual ambiguity of a moving plane. Proc R Soc Lond Ser B 223(1231):165–175
123. Longuet-Higgins HC (1986) The reconstruction of a plane surface from two perspective projections. Proc R Soc Lond Ser B 227(1249):399–410
124. Zhang Z, Hanson AR (1995) Scaled Euclidean 3D reconstruction based on externally uncalibrated cameras. IEEE international symposium on computer vision, Coral Gables, Florida, pp 37–42, Nov 1995
125. Ma Y, Soatto S et al (2003) An invitation to 3D vision, from images to geometric models. Springer, New York (ISBN 0387008934, Nov 2003)
126. Habed A, Boufama B (2006) Camera self-calibration from bivariate polynomial equations and the coplanarity constraint. Image Vis Comput 24(5):498–514
127. Sturm P (2000) Algorithms for plane-based pose estimation. Proceedings IEEE conference on computer vision and pattern recognition (CVPR00), pp 1010–1017, June 2000
128. Chum O, Werner T, Matas J (2005) Two-view geometry estimation unaffected by a dominant plane. Proceedings IEEE conference on computer vision and pattern recognition (CVPR05), pp 772–779
129. Bartoli A (2005) The geometry of dynamic scenes: on coplanar and convergent linear motions embedded in 3D static scenes. Comput Vis Image Underst 98(2):223–238 (May 2005)
130. Raij A, Pollefeys M (2004) Auto-calibration of multi-projector display walls. Proceedings of the 17th IEEE international conference on pattern recognition (ICPR'04), Washington, DC, USA, vol 1, pp 14–17, 23–26 Aug 2004
131. Okatani T, Deguchi K (2003) Autocalibration of a projector-screen-camera system: theory and algorithm for screen-to-camera homography estimation. Proceedings of the 9th IEEE international conference on computer vision (ICCV'03), vol 2, pp 774–781, Oct 2003
132. Okatani T, Deguchi K (2005) Autocalibration of a projector-camera system. IEEE Trans Pattern Anal Mach Intell 27(12):1845–1855
133. Li YF, Chen SY (2003) Automatic recalibration of a structured light vision system. IEEE Trans Rob Autom 19(2):259–268 (Apr 2003)
134. Chen SY, Li YF (2003) Self-recalibration of a color-encoded light system for automated three-dimensional measurements, Measurement Sci Tech 14(1):33–40
135. Xie Z, Zhang C, Zhang Q (2004) A simplified method for the extrinsic calibration of structured-light sensors using a single-ball target. Int J Mach Tools Manuf 44(11):1197–1203 (Sept 2004)
136. Tang A, Ng T et al (2006) Projective reconstruction from line-correspondences in multiple uncalibrated images. Pattern Recognit 39(5):889–896

137. Semple J, Kneebone G (1979) Algebraic projective geometry, Oxford University Press, Oxford
138. Shirai Y, Suwa M (1971) Recognition of polyhedrons with a range finder. Proceedings on 2nd international joint conference on artificial intelligence, pp 80–87, Sept 1971
139. Shirai Y, Tsuji S (1972) Extraction of the line drawing of 3D objects by sequential illumination from several directions. Pattern Recognit 4:343–351
140. Posdamer J, Altschuler M (1982) Surface measurement by space-encoded projected beam systems. Comput Graphi Image Process 18(1):1–17
141. Inokuchi S, Sato K, Matsuda F (1984) Range imaging system for 3D object recognition. Proceedings of the international conference on pattern recognition 1984, pp 806–808
142. Rocchini C, Cignoni P, Montani C, Pingi P, Scopigno R (2001) A low cost 3D scanner based on structured light. EG 2001 proceedings vol 20, No. 3. Blackwell Publishing, Oxford, pp 299–308
143. Petriu EM, Sakr Z, Spoelder HJW, Moica A (2000) Object recognition using pseudo-random color encoded structured light. Proceedings of the 17th IEEE conference on instrumentation and measurement technology, vol 3, pp 1237–1241
144. Zhang L, Curless B, Seitz SM (2002) Rapid shape acquisition using color structured light and multi-pass dynamic programming. International symposium on 3D data processing visualization and transmission, Padova, Italy, pp 24–26
145. Spoelder H, Vos F et al (2000) Some aspects of pseudo random binary array-based surface characterization. IEEE Trans Instrum Meas 49(6):1331–1336
146. Morano RA, Ozturk C et al (1998) Structured light using pseudorandom codes. IEEE Trans Pattern Anal Mach Intell 20(3):322–327
147. Griffin PM, Narasimhan LS, Yee SR (1992) Generation of uniquely encoded light patterns for range data acquisition. Pattern Recognit 25(6):609–616
148. Griffin PM, Yee SR (1996) A decoding algorithm for unique correspondence in range data acquisition. Int J Prod Res 34(9):2489–2498
149. Salvi J, Pages J, Batlle J (2004) Pattern codification strategies in structured light systems. Pattern Recognit 37(4):827–849
150. Hartley R, Zisserman A (2000) Multiple-view geometry in computer vision. Cambridge University Press, New York
151. Li M (1994) Camera calibration of a head-eye system for active vision. Proceedings of the European conference on computer vision 1994, pp 543–554
152. Lens RK, Tsai RY (1988) Techniques for calibration of the scale factor and image center for high accuracy 3D machine vision metrology. IEEE Trans Pattern Anal Mach Intell 10(5):713–720 (Sept 1988)
153. Sturm P, Cheng Z et al (2005) Focal length calibration from two views: method and analysis of singular cases. Comput Vis Image Underst 99(1):58–95
154. Cao X, Xiao J et al (2006) Self-calibration from turn-table sequences in presence of zoom and focus. Comput Vis Image Underst 102:227–237
155. Kanatani K, Nakatsuji A, Sugaya Y (2006) Stabilizing the focal length computation for 3D reconstruction from two uncalbrated views. Int J Comput Vison 66(2):109–122
156. Borghese N, Colombo M, Alzati A (2006) Computing camera focal length by zooming a single point. Pattern Recognit 39(8):1522–1529
157. Raskar R, Beardsley P (2001) A self correcting projector. Proceedings IEEE conference on computer vision and pattern recognition (CVPR01), Hawaii, Dec 2001
158. Pavlidis T (1982) Algorithms for graphics and image processing. Computer science press. Springer, Berlin
159. Huynh DQ (2000) The cross ratio: a revisit to its probability density function. British machine vision conference, Bristol, United Kingdom, vol 1, pp 262–271, 11–14 Sept 2000
160. Negahdaripour S (1990) Closed form relationship between the two interpretations of a moving plane. J Opt Soc Am 7(2):279–285
161. Negahdaripour S (1990) Multiple interpretations of the shape and motion of objects from two perspective images. IEEE Trans Pattern Anal Mach Intell 12(11):1025–1039
162. Hartley R (1998) Chirality. Int J Comput Vosion 26(1):41–61

163. Maybank S (1992) Theory of reconstruction from image motion. Springer, New York
164. Tian T, Tomasi C, Heeger D (1996) Comparison of approaches to egomotion computation. Proceedings of the IEEE conference on computer vision and pattern recognition, San Francisco, pp 315–320, June 1996
165. Forsyth D, Ponce J (2003) Computer vision: a modern approach. Prentice Hall, London
166. Criminisi A, Reid I, Zisserman A (1999) A plane measuring device. Image Vis Comput 17:625–634
167. Grossmann E, Santos-Victor J (2000) Uncertainty analysis of 3D reconstruction from uncalibrated views. Image Vis Comput 18:685–696
168. Sun Z, Ramesh V, Tekalp AM (2001) Error characterization of the factorization method. Comput Vis Image Underst 82(2):110–137
169. Wang L, Kang SB et al (2004) Error analysis of pure rotation-based self-calibration. IEEE Trans Pattern Anal Mach Intell 26(2):275–280
170. Weng J, Huang T, Ahuja N (1989) Motion and structure from two perspective views: algorithms, error analysis and error estimation. IEEE Trans Pattern Anal Mach Intell 11(5):451–476
171. Wiley AG, Wong KW (1995) Geometric calibration of zoom lenses for computer vision metrology. Photogramm Eng Remote Sens 61(1):69–74
172. Fofi D, Salvi J, Mouaddib E (2003) Uncalibrated reconstruction: an adaptation to structured light vision. Pattern Recognit 36(7):1631–1644
173. Ponce J (2009) What is a camera? IEEE international conference on computer vision and pattern recognition, pp 1526–1533, 20–25 June 2009
174. Han M, Kanade T (2000) Scene reconstruction from uncalibrated views, CMU-RI-TR-00–09, Jan 2000
175. Zhang Z (2004) Camera calibration with one-dimensional objects. IEEE Trans Pattern Anal Mach Intell 26(7):892–899
176. Zhao Z, Wei Z, Zhang G (2009) Estimation of projected circle centers from array circles and its application in camera calibration. The 2nd Asia-Pacific conference on computational intelligence and industrial applications 2009, pp 182–185
177. Jiang G, Quan L (2005) Detection of concentric circles for camera calibration. Proceedings IEEE international conference on computer vision, vol 1, pp 333–340
178. Micusik B, Pajdla T (2006) Structure from motion with wide circular field of view cameras. IEEE Trans Pattern Anal Mach Intell 28(7):1135–1149
179. Berhanu M (2005) The polynomial eigenvalue problem. PhD thesis, University of Manchester
180. Kukelova Z, Pajdla T (2007) A minimal solution to the autocalibration of radial distortion. Proceedings of computer vision and pattern recognition conference, 2007, pp 1–7
181. Byrod M, Kukelova Z, Josephson K et al (2008) Fast and robust numerical solutions to minimal problems for cameras with radial distortion. Proceedings of computer vision and pattern recognition conference, 2008, pp 1–8
182. Devernay F, Faugeras O (2001) Straight lines have to be straight. Mach Vis Appl 13:14–24
183. Ahmed M, Farag A (2005) Nonmetric calibration of camera lens distortion: differential methods and robust estimation. IEEE Trans Image Process 14(8):1215–1230
184. Ma SD, Zhang Z (1998) Computer vision: computational theory and algorithms. The Science Press of China, Beijing
185. Zhang B, Li YF, Fuchao W (2003) Planar pattern for automatic camera calibration. Opt Eng 42(6):1542–1549
186. Zhang B, Dong L, Wu FC (2001) Camera self-calibration based on circle. Proc SPIE Vis Optim Tech 4553(9):286–289
187. Lhuillier M (2007) Toward flexible 3D modeling using a catadioptric camera. IEEE international conference on computer vision and pattern recognition, pp 1–8, June 2007
188. Ragot N, Rossi R et al (2008) 3D volumetric reconstruction with a catadioptric stereovision sensor. IEEE Int Symp Ind Electron 11:1306–1311
189. Boutteau R, Savatier X et al (2008) An omnidirectinal stereoscopic system for mobile robot navigation. IEEE Int Workshop Rob Sensors Environ 17–18(10):138–143

190. Micusik B, Pajdla T (2006) Structure from motion with wide circular field of view cameras. IEEE Trans Pattern Anal Mach Intell 28(7):1135–1149
191. Baker S, Nayar SK (1998) A theory of catadioptric image formation. Proceedings of the international conference on computer vision, pp 35–42, Jan 1998
192. Geyer C, Daniilidis K (1999) Catadioptric camera calibration. IEEE proceedings of the 7th international conference on computer vision, vol 1, pp 398–404, 20–27, Sept 1999
193. Barreto P, Araujo H (2001) Issues on the geometry of central catadioptric image information. IEEE conference on computer vision and pattern recognition, 2001, pp 422–427
194. Mei C, Malis E (2006) Fast central catadioptric line extraction, estimation, tracking and structure from motion. IEEE/RSJ international conference on intelligent robots and systems, pp 4774–4779, Oct 2006
195. Geyer C, Daniilidis K (2002) Paracatadioptric camera calibration. IEEE Trans Pattern Anal Mach Intell 24(5):687–695
196. Barreto P, Araujo H (2005) Geometric properties of central catadioptric line images and their application in calibration. IEEE Trans Pattern Anal Mach Intell 27(8):1327–1333
197. Orghidan R, Mouaddib E, Salvi J (2005) Omnidirectional depth computation from a single image. IEEE proceedings of international conference on robotics and automation, pp 1222–1227, Apr 2005
198. Baldwin J, Basu A (2000) 3D estimation using panoramic stereo. IEEE proceedings of international conference on pattern recognition, vol 1, pp 97–100, Sept 2000
199. Aliega G (2001) Accurate catadioptric calibration for real-time pose estimation in room-size environments. IEEE proceedings of the 9th international conference on computer vision, vol 1, pp 127–134, July 2001
200. Svoboda T, Padjla T, Hlavac V (2002) Epipolar geometry for panoramic cameras. Int J Comput Vis 49(1):23–37
201. Barreto J, Daniilidis K (2006) Epipolar geometry of central projection systems using veronese maps. IEEE proceedings of the conference on computer vision and pattern recognition, vol 1, pp 1258–1265, June 2006
202. Micusik B, Pajdla T (2003) Estimation of omnidirectional camera model from epipolar geometry. IEEE proceedings of the conference on computer vision and pattern recognition, vol 1, pp 485–490, June 2003
203. Mei C, Benhimane S, Malis E, Rives P (2006) Homography-based tracking for central catadioptric cameras. IEEE/RSJ international conference on intelligent robots and systems, pp 669–674, Oct 2006
204. Lhuillier M (2008) Automatic scene structure and camera motion using a catadioptric system. Comput Vis Image Underst 109(2):186–203
205. Goncalves N, Araujo H (2009) Estimating parameters of non-central catadioptric systems using bundle adjustment. Comput Vis Image Underst 113:186–203
206. Toepfer C, Ehlgen T (2007) A unifying omnidirectional camera model and its applications. Proceedings of the 11th IEEE international conference on computer vision, 2007
207. Grossberg MD, Nayar SK (2005) The raxel imaging model and ray-based calibration. Int J Comput Vis 61:119–137
208. Tardif JP, Sturm P, Trudeau M, Roy S (2009) Calibration of cameras with radially symmetric distortion. IEEE Trans Pattern Anal Mach Intell 31(9):1552–1566
209. Baker S, Nayar SK (1999) A theory of single-viewpoint catadioptric image formation. Int J Comput Vis 35(2):175–196
210. Geyer C, Daniilidis K (2001) Catadioptric projective geometry. Int J Comput Vis 45(3):223–243
211. Barreto P (2006) A unifying geometric representation for central projection systems. Comput Vis Image Underst 103(3):208–217 (Sept 2006)
212. Ying X, Hu Z (2004) Can we consider central catadioptric cameras and fisheye cameras within a unified imaging model. Proceedings European conference on computer vision (ECCV04), vol 1, pp 442–455

213. Li W, Li YF, Sun D, Li Z, Zhang B (2011) Generic radial distortion calibration of a novel single camera based panoramic stereoscopic system. IEEE international conference on robotics and automation Shanghai, China, 9–13 May, 2011
214. Astre B, Sarry L et al (2008) Automatic calibration of a single-projector catadioptric display system. IEEE international conference on computer vision and pattern recognition, pp 1–8
215. Ding Y, Xiao J et al (2009) Catadioptric projectors. IEEE international conference on computer vision and pattern recognition, pp 2528–2535